Neurobiology of "Umwelt"

RESEARCH AND PERSPECTIVES IN NEUROSCIENCES

Fondation Ipsen

A. Berthoz • Y. Christen
Editors

Neurobiology of "Umwelt"

How Living Beings Perceive the World

Springer

Editors

Berthoz, Alain
CNRS UMR 9950 Collège de France
11 Place Marcelin Berthelot
75231 Paris Cedex 05
France
alain.berthoz@college-de-france.fr

Christen, Yves
Fondation IPSEN
Pour la Recherche Thérapeutique
65, quai georges Gorse
92650 Boulogme Billancourt
Cedex-France
yves.christen@ipsen.com

ISSN: 0945-6082
ISBN: 978-3-540-85896-6 e-ISBN: 978-3-540-85897-3

Library of Congress Control Number: 2008934399

Printed on acid-free paper

springer.com

Foreword

At the beginning of the 20th century, German biologist Jakob von Uexküll created the concept of *Umwelt* to denote the environment as experienced by a subject. This concept of environment differs from the idea of passive surroundings and is defined not just by physical surroundings, but rather is a "subjective universe", a space weighted with meaning. Based on this perspective, a living organism, no matter how basic (such as the tick studied by von Uexküll), creates its own universe when it interacts with the world and as this same time the organism reshapes it. Today, neuroscience provides a new way to look at the brain's capability to create a representation of the world. At the same time, behavioral specialists are demonstrating that animals have a richer mental universe than previously known. Philosophical reflection thus finds itself with more experimental and objective data as well. This is why we have chosen the theme of *Umwelt,* nearly a century after the publication of von Uexküll's founding work (*Umwelt and Innenwelt der Tiere* was published in 1909), for the 16th international "Colloque Médecine et Recherche" in neuroscience organized by the Fondation Ipsen. This meeting bring together neurobiologists, psychologists, sociologists, anthropologists, ethologists, and philosophers, in Paris on February 18, 2008.

Alain Berthoz
Yves Christen

Acknowledgments

The editors wish to express their gratitude to Mrs Mary Lynn Gage for her editiorial assistance, Mrs Sonia Le Cornec and Jacqueline Mervaillie for the organization of the meeting.

Contents

Contributors

Berthoz Alain
CNRS UMR 9950 Collège de France, 11 Place Marcelin Berthelot, 75231 Paris Cedex 05, France, Alain.berthoz@college-de-France.fr

Cheney Dorothy L.
University of Pennsylvania, Departments of Biology & Psychology, 3720 Walnut St. Room D7, Philadelphia, PA 19104, USA

Clayton Nicola S.
Department of Experimental Psychology, University of Cambridge, Downing Street, Cambridge CB2 3EB, UK, nsc22@cam.ac.uk

Emery Nathan J.
School of Biological & Chemical Sciences, Queen Mary, University of London, London E1 3NS, UK

Fagot-Largeault Anne
College de France, Philosophy of Life Science, 11 place Marcelin Berthelot, 75231 Paris Cedex 05, France, anne.fagot-largeault@college-de-france.fr

Gelman Susan A.
Department of Psychology, University of Michigan, 525 East University Avenue, Ann Arbor, MI 48109-1109, USA, gelman@umich.edu

Hare Brian
Department of Biological Anthropology and Anatomy, Duke University, Durham NC 27708, USA, +1-919-660-7292, b.hare@duke.edu

Holland Richard
Marie Curie Outgoing International Fellow, Institute for Integrative and Comparative Biology, University of Leeds, Leeds, LS2 9JT, UK, bgyraho@leeds.ac.uk

Ingold Tim
Department of Anthropology, School of Social Science, University of Aberdeen, F 50 Dunbar Street, Aberdeen AB24 3QY, Scotland, UK, tim.ingold@abdn.ac.uk

Jacobs Gerald H.
Neuroscience Research Institute, University of California, Santa Barbara, CA 93106, USA, jacobs@psych.ucsb.edu

Kaminski Juliane
Sub-Department of Animal Behaviour, University of Cambridge, High Street, Madingley, Cambridge CB3 8AA, UK, Kaminski@eva.mpg.de

Llinàs Rodolfo R.
Department of Physiology and Neuroscience, New York University Medical School, 550 First Avenue, New York, NY 10016, USA, llinar01@popmail.med.nyu.edu

Seyfarth Robert M.
University of Pennsylvania, Departments of Biology & Psychology, 3720 Walnut St. Room D7, Philadelphia, PA 19104, USA, seyfarth@psych.upenn.edu

Singer Wolf
Department of Neurophysiology, Max Planck Institute for Brain Research, Deutschordenstrasse 46, 60528 Frankfurt am Main, Germany
singer@mpih-frankfurt.mpg.de

Anthropological Physiology: von Uexküll, Portmann, Buytendijk

Anne Fagot-Largeault

Abstract The notion of Umwelt originates in a (continental, 'anthropological') tradition of studying the behaviour of animals in their natural environment, contrasting with the (anglo-saxon, behaviorist) tradition of breeding mice in the laboratory and testing their achievements in mazes, that is, in artificial environments. Chapter 1 outlines the contributions of three major European scientists to modern psychophysiology and ethology.

"Philosophical anthropology" refers to a trend of thought that flourished around the middle of the 20th century (between 1920 and 1960) on the European continent, especially in Germany, Switzerland and the Netherlands. It comes as a humanistic reaction against positivistic naturalism in science, especially the new experimental sciences of human and animal behaviour. It strives to build a bridge between *Naturwissenschaften* and *Geisteswissenschaften*, that is, between the science of nature and the science of the human mind. The basic concern is to initiate a philosophical way of practicing science. It aims at understanding how the human race "builds its nest" in the world (Gehlen, posth., 1986). The philosophical inspiration lies both in existentialism (Karl Jaspers) and phenomenology (Edmund Husserl). Most of the literature is in German. *Ethology* and *physiological anthropology* belong to that trend of thought.

After world war 2, continental biologists who had emigrated to England reckoned that the study of animal behaviour had developed along so divergent paths in the anglophone world and in the german world, that researchers did not understand each other any more. They had published in different journals, writing different languages, using different technical vocabulary, and different methods of research and measurement. On the Anglo-american side, the research had mainly been the job of psychologists, the typical animal was the laboratory mouse (or rat), the main focus was learning, the explanatory scheme was Pavlov's conditioning. That was the *behaviorist* school. On the German side, research had been conducted mainly

A. Fagot-Largeault
Collège de France, Philosophy of Life Science, 11 place Marcelin Berthelot 75005 Paris, France
e-mail: anne.fagot-largeault@college-de-france.fr

A. Berthoz and Y. Christen (eds.), *Neurobiology of "Umwelt": How Living Beings Perceive the World,* Research and Perspectives in Neurosciences,

by zoologists. Animals - even humble animals such as insects - had been studied outdoor, within their natural environment; learning was not the focus of interest; German researchers were interested in innate behaviour, instinctive (often complex) animal reproductive (or other) strategies. That was *ethology*. In 1950 a meeting took place in Cambridge, England, during which researchers of the two traditions met and discussed with each other. Paul Schiller, a psychologist from Hungary, offered to translate the German literature in English. He died before the work was done, but his wife completed it (Coll., *Instinctive Behaviour*, 1957). That is when the two schools merged, and when, in animal research, the Darwinian scheme of explanation (*trial and error*) definitely replaced the behaviorist scheme (*conditioning*).

"In the behaviorist's Umwelt the body produces the mind, and in the psychologist's world the mind builds the body" (von Uexküll 1934, p. 80).

The promoter of German ethology was **Jakob von Uexküll** (1864–1944), who says that he took his inspiration from Johannes Müller, the initiator in Germany of *physiology* as a science. Note that there has been two von Uexküll working in the field : the father Jakob, who was born in Estonia, studied zoology in Tartu, and later did most of his research in Heidelberg and the Zoological Institute in Rostock; and his son Thure von Uexküll (1908–2004), also a partner of the philosophical anthropology movement.

The famous description, by Jakob von Uexküll, of the phenomenal world of the tick (Uexküll 1921), is meant to convey the idea that even such a humble animal is not only "a collection of perceptual and effector tools connected by an apparatus which mechanically … carries on the life functions", but that there is a pilot in the machine, *i.e.*, a *subject*, "whose essential activity consists in perceiving and acting". The functional cycle is very simple. To carry it out (she can wait 18 years) she has three receptors (a photosensitive skin, a sense of the smell of butyric acid, a sense of warm temperature) and three effectors (climb up the tip of a branch, drop on a passing mammal, drink its warm blood). Then the cycle is over: "nothing left for her to do but drop to earth, lay her eggs and die" (p. 7). What is the tick's *Umwelt*? "Perceptual and effector worlds together form a closed unit, the Umwelt", says von Uexküll (1934, in: Coll., p. 6). And he goes on: "Now we might assume that an animal is nothing but a collection of perceptual and effector tools, connected by an integrating apparatus which, though still a mechanism, is yet fit to carry on the life functions. This is indeed the position of all mechanistic theorists, whether their analogies are in terms of rigid mechanics or more plastic dynamics. They brand animals as mere objects. The proponents of such theories forget that, from the first, they have overlooked the most important thing, the *subject* which uses the tools, perceives and functions with their aid" (J. von Uexküll, 1934, in: Coll., p. 6).

What the author means is that the animal actively builds her Umwelt, and that such a construction reveals a living strategy. "As the spider spins its threads, every subject spins his relations to certain characters of the things around him, and weaves them into a firm web which carries his existence" (J. von Uexküll, 1934, p. 14). While weaving their niche, living beings prepare themselves to be responsive to certains cues in the world around them, and even though their behaviour may not be consciously planned, it is obviously meaningful, to the extent that it serves a survival or reproductive end. The author distinguishes between innate and learned behaviour, but neither behaviour is mechanically triggered off by the environment. "If we choose to call significant only what is given to the subject by the evidence of his senses, then, of course, only the familiar path will be called meaningful, not the innate. Even so, it remains planful to the highest degree" (J. von Uexküll, 1934, 70).

There is a counterpart to such an analysis of the living being as a perceiving-and-acting subject. The subject is locked in her niche. For the tick there is nothing beyond the edge of her Umwelt; for her, all mammals reduce to some thing with a smell of butyric acid. Von Uexküll invites the reader to imagine an oak tree, with squirrels running and birds nested on its branches, a fox living between its roots, a bunch of beetles on its bark, each with their own Umwelt, and an ant. "Each Umwelt carves a specific section out of the oak, whose qualities are suitable bearers for both the receptor and effector cues of their respective functional cycles. In the ant's world all the rest of the oak vanishes behind its gnarled bark, whose furrows and heights become the ant's hunting ground" (von Uexküll, 1934, p. 75).

Adolf Portmann (1897–1982) was a zoologist known for having developed the idea that human beings were born *premature*, and that the extra-uterine embryos we all were, found a *second uterus* in their social environment. Biological development in interaction with a human milieu offers a possibility of vast diversification. Understanding biological development requires both an analysis of particular developmental mechanisms, and a holistic view of organic structures and strategies : "Das lebendige Geschehen zeigt in jedem Ausschnitt den Doppelaspekt, der einerseits die Untersuchung dienender Strukturen und Wirkweisen erfordert, und der anderseits verlangt, dass wir zugleich um das übergeordnete Ganze wissen, das diese dienenden Strukturen benützt" (Portmann 1951, p. 90). In other words, biologists when studying human embryological and postnatal development cannot ignore the regulating influence of the psychological and social 'Umwelt' on physical development, just like psychologists when studying the mind cannot ignore that the mind is *embodied*, that is, enveloped in a totality: the body. In his book on *The Animal as a Social Being* (*Das Tier als soziales Wesen*, Chap. 1), Portmann starts with a minute description of the *Libellenwelt* (the world of a dragonfly).

Note that the idea of the social milieu being a second uterus had already be expressed by Antoine Augustin Cournot (in his book: *Mechanism, Vitalism, Rationalism,* 1875, §8). Cournot mentions what must have been a lecture by Claude Bernard, as a reference for such an idea. Portmann does not mention either Cournot or Bernard. He seems to have developed the idea independently.

Frederik Jacobus Johannes **Buytendijk** (1887–1974) is the main scholar representative of anthropological physiology in the Netherlands. He was not a zoologist. After studying medicine in Amsterdam, he went into experimental physiology, and became an expert in animal behaviour, or, to be precise, in animal psycho-physiology. He radically disagreed with John Watson's mechanistic model of conditioning, and with the ways of experimenting on animals common in the behaviourist school. His inaugural address on "understanding living phenomena" when, in 1925, he was established as professor of physiology at the university of Groningen, makes it explicit that his research programme consists in studying animals and/or humans as "psycho-somatic units" in other than artificial environments. From then on he concentrated more and more on psychophysiological phenomena which may be qualified as "modes of being", such as pain, fatigue, hunger, thirst, anger and other emotional states, and on organic regulations (for example, of blood pressure, or equilibrium). In so doing he used both cybernetic models and notions borrowed from phenomenology, especially the notion of an intentionality inherent in behaviour. Buytendijk assumes that there is in animals a living *subjectivity*, and that subjectivity does not necessarily imply *consciousness*: "The concept of a lived subjectivity which is bodily unconscious is in accordance with the experiences we have of the behaviour of animals" (*Prolegomena*, A, I, § 6).

Methodologically speaking, Buytendijk's main thesis is that the living being, which is the *object* of scientific study, is at the same time another *subject*, and that he must be treated as such, that is, not be experimented on, but be a partner in experimental research. One does not do research *on* an ape, one does research *with* (the cooperation of) an ape. The other subject does not react to *stimuli*, he reacts to the *meaning* the stimuli have for him. The experiment is an *encounter*. Buytendijk's ideas on experimental research have been most influential on the ethics of medical research during the late twentieth Century. Buytendijk himself is clearly aware of the potential impact on medicine of his concept of a *research partnership*, when he compliments the French psychiatrist Henry Ey for translating German works into French, and understanding that, in order to do good science, one does not need to reify (that is, treat as a thing) the object of research : "*The reintroduction of the subject into physiology and biology is the chief concern of modern thought*' - The import of this statement by the psychiatrist Henri Ey - in his introduction to the French translation of *Der Gestaltkreis*, the pioneering work of von Weizsäcker - can only be understood if we withdraw from the activity of technically-oriented medical science" (*Prolegomena*, A, I, § 7).

Buytendijk's *Prolegomena to an anthropological physiology* is available in English. Buytendijk was a wonderfully learned researcher, speaking and writing several languages. When he turned seventy, colleagues from all over Europe composed a book in his honor, in three languages, under the title of "Rencontre, Encounter, Begegnung". Maurice Merleau-Ponty (1908–1961) borrowed much of his empirical data from Buytendijk (see Dekkers' paper, 1995). Buytendijk generously expressed admiration for Merleau's theory of the human body as a mode of 'being-in-the-world'. It seems, however, that Buytendijk's notion of *Leiblichkeit* (bodiliness) goes further than Merleau's: "*Spirit manifests itself through the body*' - the basic thesis of Merleau-Ponty. We add to this: 'The body of man organizes itself in its human performances and structurations through the mind" (*Prolegomena*, A, I, § 7).

References

Buytendijk Frederik J.J., *Über den Schmerz*, aus dem Holländischen übersetzt, Huber, 1948; *De la douleur*, tr. fr. d'après la version allemande A. Reiss, préf. Maurice Pradines, Paris: PUF, 1951.

Buytendijk F.J.J., 'Zur Phänomenologie der Begegnung', *Eranos-Jahrbuch*, 1951 (19): 431–486; tr. fr. J. Knapp, *Phénoménologie de la rencontre*, Paris: Desclée, de Brouwer & Cie, 1952.

Buytendijk F.J.J., *Prolegomena van een anthropologische fysiologie*, 1965; German tr., *Prolegomena einer anthropologichen Physiologie*, Salzburg: Otto Müller, 1967, 307 p; Engl. tr. *Prolegomena to an Anthropological Physiology*, Pittsburgh: Duquesne University Press, 1975.

Coll., *Instinctive Behavior. The development of a modern concept*, transl. & ed. Claire H. Schiller, intr. Karl S. Lashley, contrib. D.J. Kuenen, Konrad Lorenz, Nicholas Tinbergen, Paul H. Schiller, Jakob von Uexküll, London: Methuen, 1957.

Coll., *Rencontre Encounter Begegnung. Contributions à une psychologie humaine dédiées au Professeur F.J.J. Buytendijk*, Utrecht: Uitgeverij het Spectrum, 1957, 520 p.

Dekkers Wim J.M., 'F.J.J. Buytendijk's concept of an anthropological physiology', *Theoretical Medicine*, 1995, 16: 15–39.

Gehlen Arnold, *Anthropologische und sozialpsychologische Untersuchungen*, Reinbek bei Hamburg: Rowohlt Taschenbuch Verlag, 1986, posth.; tr. fr. *Anthropologie et psychologie sociale*, Paris: PUF, 1990.

Portmann Adolf, *Die Biologie und das neue Menschenbild*, Bern, 1942; *Zoologie und das neue Bild des Menschen*, Hamburg, 1956.

Portmann A., *Das Tier als soziales Wesen*, Zürich: Rhein-Verlag, 1953, 2e ed. 1962, 382 p.

Portmann A., 'Über die Eigenart des biologischen Forschens', in: Grassi E. & Uexküll Thure von, eds., *Die Einheit unseres Wirklichkeitsbildes und die Grenzen der Einzelwissenschaften*, Bern: Franke, 1951, 69–93.

Uexküll Jakob von, *Umwelt und Innenwelt der Tiere*, Berlin, 1921.

Uexküll J. von, *Streifzüge durch die Umwelten von Tieren und Menschen*, Berlin: Springer, 1934; Engl. tr. C.H. Schiller, 'A stroll through the worlds of animals and men. A picture book of invisible worlds', in: Coll., *Instinctive Behavior*, 1957.

Uexküll J. von, *Der Sinn des Lebens*, Stuttgart: Ernst Klett, 1977; herausgegeben von F.G. Jünger & M. Himmelhaber, *Scheidewege. Vierteljahresschrift für skeptisches Denken*, Beiheft 4, 1977.

Weizsäcker Viktor von, *Der Gestaltkreis: Theorie der Einheit von Wahrnehmen und Bewegen*, Leipzig: G. Thieme, 1939, 2nd ed. 1943, 3rd ed. 1946, 4th ed. 1948, (in: *Gesammelte Schriften*, Bd. 4); Fr tr. by M. Foucault & D. Rocher, Pref. Henri Ey, *Le cycle de la structure*, Paris: Desclée, 1958.

Essentialist Reasoning about the Biological World

Susan A. Gelman

Abstract Essentialism is the idea that certain categories, such as "dog," "man," or "gold," have an underlying reality or true nature that gives objects their identity. Essentialist accounts have been offered, in one form or another, for thousands of years, extending back at least to Aristotle and Plato. Where does this idea come from? I address this question from a psychological perspective and argue that essentialism is an early cognitive bias. Young children's concepts reflect a deep commitment to essentialism, and this commitment leads children to look beyond the obvious in many converging ways: when learning language, generalizing knowledge to new category members, reasoning about the insides of things, contemplating the role of nature versus nurture, and constructing causal explanations. I suggest that children have an early, powerful tendency to search for hidden, non-obvious features of things. Parents do not explicitly teach children to essentialize; instead, during the preschool years, children spontaneously construct concepts and beliefs that reflect an essentialist bias. I explore the broader implications of this perspective for human concepts, children's thinking, and the relation between human concepts and the biological world.

1 Introduction

One important task that humans face as they experience the biological world is to organize it into categories. Categorization serves two important functions: it provides an efficient system for storing the endless variety of sights, sounds, and events that we encounter, and it provides a structure for making new inferences and predictions (Smith 1989). All animals use categories in these ways. Detecting food, enemies, or prey all require responding to new and perceptibly distinct items as if they were comparable to previously viewed items.

S.A. Gelman
Department of Psychology, University of Michigan, Ann Arbor, MI
e-mail: gelman@umich.edu

A. Berthoz and Y. Christen (eds.), *Neurobiology of "Umwelt": How Living Beings Perceive the World,* Research and Perspectives in Neurosciences,

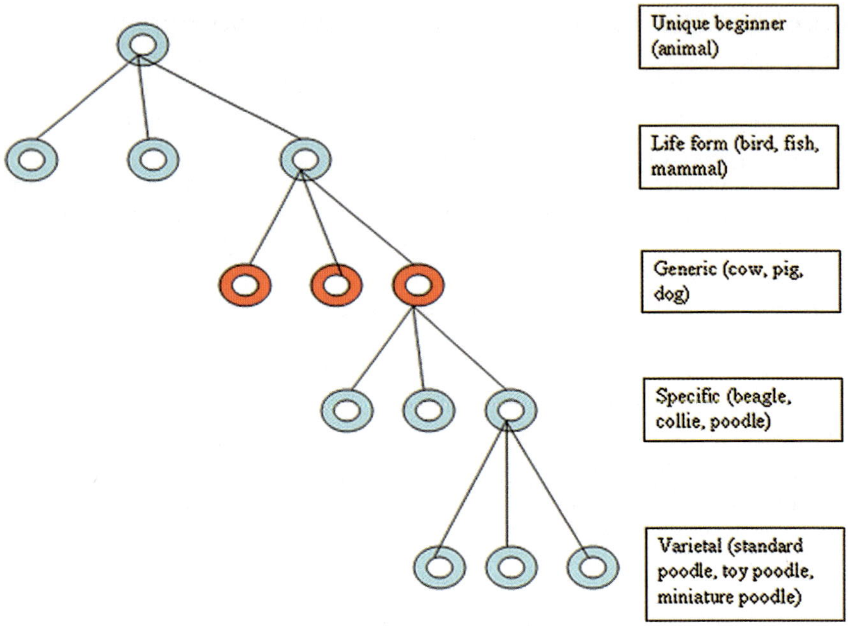

Fig. 1 Sample partial folk taxonomy. Note: each circle represents a category. The levels (with examples) are listed in boxes to the right

Although all organisms form categories, only humans further organize categories into taxonomic systems. These taxonomies, expressed universally in human languages, have several features: they include three to five levels, categories are mutually exclusive within a level, and categories at one level are nested within the next-highest level (Berlin 1992; Atran 1990). See Figure 1 for a partial example. Furthermore, one of the most notable aspects of these "folk taxonomies" is that, universally, the middle level (typically the generic, though sometimes specific; e.g., dog, sheep, lion) seems to be privileged. This level is most typically named with a single word (Berlin 1992), it is the earliest acquired in childhood (Mervis and Rosch 1981), and it is most often used as the basis for making inferences about novel properties (Coley et al. 1999). In short, humans the world over seem to share a taxonomic impulse: a powerful propensity to codify and systematize the natural world.

At the same time that people behave like untutored systematists, we also are prone to some surprising errors in reasoning about biological categories. I give three examples here. First, people are surprisingly resistant to evolutionary theory. Nearly half the U.S. population rejects evolutionary theory altogether (Evans 2001). Even among those who endorse evolution and have passed college-level biology classes, many misinterpret evolutionary theory in fundamental ways, reporting that evolutionary changes occur not at the population level but rather at the level of individual organisms changing to fit into the environment (Shtulman 2006). Second, people misinterpret the concept of genes. For example, most U.S. adults agree with the erroneous statement, "Two people from the same race will always be more genetically similar to each other than two people from different races" (Jayaratne 2001). They do not appreciate that within-category variation extends to the level of non-obvious underlying features. Instead, they seem to believe that all members of a given category (e.g., Africans, Europeans, Asians) are unvarying at an internal level. Third, people believe that internal bodily organs (such as hearts or lungs) can affect a person's behavior or personality. Over one-third of heart-transplant recipients believe that they take on characteristics of the donor (Inspector et al. 2004), in some cases reporting, for example, that a male heart transplanted into a woman's body will cause the woman to behave more like a man (Sylvia and Novak 1997).

A common framework that can unite these varied misconceptions is that of "psychological essentialism." Philosophers have long proposed that certain categories are supported by "essences," an underlying reality or true nature that one cannot observe directly but that gives an object its identity (Schwartz 1977). This metaphysical framework is problematic from a biological perspective (Dupré 1993; Sober 1994; Wilson 1999). However, in everyday thought people seem to endorse an essentialist view of the natural world (Gelman 2003; see Strevens 2000, Ahn et al. 2001, for discussion). They treat natural categories (e.g., tigers, females, gold) as real and unchanging, and further believe that there is some unobservable, underlying quality (i.e., the essence) that all members of a category share and that causes observable characteristics. The essence may be some inner part or substance (e.g., DNA, blood), or it may be unknown. Medin and Ortony (1989) refer to this as an essence placeholder: the notion that there is some underlying cause, as yet to be discovered.

A variety of studies and methods have obtained evidence for essentialist reasoning among adults (Haslam and Ernst 2002; Prentice and Miller 2007; Medin and Ortony 1989; Ahn et al. 2006; Rehder 2007). At times people refer to essentialism explicitly, as in the quote below regarding a woman (Claire) who received a heart from a donor (Tim): "I am beginning to believe that some of Tim's essence has transmigrated to Claire. … If the transplant has somehow passed on elements of his temperament, personality, and identity, then psychological residues of the actual Tim L. (not just the image of 'Tim') may now inhabit Claire" (Sylvia and Novak 1997, p. 165). At other times people appeal to an essence only implicitly, as with the six-year-old who told me: "All snakes are a little bit same and a little bit different. *Inside*, they're the same." Nonetheless, essentialism seems to be widely endorsed.

A key question these data raise is where an essentialist bias comes from. There are competing theoretical accounts in the literature. One possibility is that essentialism is the product of particular sociological, political, or historical conditions (Fuss 1989; Guillaumin 1980). For example, Fodor (1998) proposed that essentialism derives from modern scientific and technological knowledge. In contrast, others have proposed that essentialism is a fundamental human bias, one that does not require particular cultural experiences but rather emerges spontaneously across many different contexts (Atran 1998; Bloom 2000; Gil-White 2001; Pinker 1994).

Young children have proven to be a particularly useful group in which to examine these competing claims. In most cultures, children below five or six years of age do not yet receive formal schooling and so have little exposure to scientific instruction regarding possible essentialist accounts (e.g., genes, DNA) or Western philosophical traditions (e.g., Plato, Aristotle). Nonetheless, converging evidence suggests implicit essentialism in young children, thus undermining the argument that essentialism requires particular knowledge of scientific advances or historical conditions. Essentialist beliefs in children (to be briefly reviewed below) support the notion that essentialism is a basic, spontaneous human bias.

How might one test for essentialism? One method that has proven productive with adults is to provide a series of survey questions that participants can endorse or reject, on a Likert scale. For example, below are listed examples of wording adapted from that used by Haslam et al. (2000) to assess the essentialist principles of immutability, innateness, and underlying reality. As is evident, these questions require a metacognitive ability to reflect on one's own beliefs.

Immutability: "It is not easy to change this category; it is fixed."

Innateness: "The kind of category something is can be largely attributed to its genetic inheritance."

Underlying reality: "Some categories have an underlying reality; although their members have similarities and differences on the surface, underneath they are basically the same."

It would be challenging to extend such questions to preschool-aged children, given their limited vocabulary and limited metacognitive skills. Instead, a series of methodologies have been developed to assess essentialism, each designed to suit the limited attentional capacities of a child under five years of age. Here I illustrate briefly with four different techniques, designed to assess immutability, inductive potential, innateness, and underlying reality, respectively (see Gelman 2003 for a more detailed review).

Immutability. An essentialist view proposes that categories are fixed and unchanging, even in the face of salient or dramatic outward changes in appearance. Although this belief can have negative implications (most notably, rejection of evolutionary theory), it also makes the appropriate prediction that an individual animal of one kind cannot be transformed into another kind (for example, a raccoon cannot become a skunk). Keil (1989) found evidence for this belief in children by about seven years of age. In his task, children saw a series of animals that underwent a series of modifications so that they appeared to change category membership. For example, a raccoon was modified by a scientist to look and seem just like a skunk (including having a smelly pouch inserted so that it could spray a smelly odor). By about seven years of age, children reported that the animal was still a raccoon, even though it looked so much like a skunk. With somewhat simpler items, even younger children (three to five years of age) showed this effect, expressing the belief that outward changes left identity unchanged (Keil 1989; Gelman and Wellman 1991). By four years of age, children treated category membership as stable and unchanging over transformations such as costumes, growth, metamorphosis, or changing environmental conditions.

Inductive potential. Inductive reasoning entails making inferences about what is not yet known. For example, if you ate a gray speckled mushroom last week without suffering ill effects, you might reasonably infer that another gray speckled mushroom will also be edible. However, what would you do when encountering another mushroom that looks like the one you ate last week but is called by a different name or one that looks different but has the same label? In other words, which is more crucial: the outward perceptual cues or the category label? A series of studies indicate that children from 13 months of age onward, like adults, make use of the category label to infer novel properties (Graham et al. 2004; Jaswal and Markman 2007; Gelman and Markman 1986). For example, children are more likely to extend a newly learned fact from a black beetle to a green leaf-insect than from a green leaf to a green leaf-insect, despite the greater similarity of the latter pair. This effect holds up for animals (bird, fish), natural substances (gold, cotton), and social categories (boy, girl, smart, shy). Young children's category-based inferences are consistent with essentialism in two respects: the properties being projected are non-obvious properties, and children can overlook outward appearances in making property projections.

Sloutsky and Fisher (2004) havc argued that this labeling effect is due to item similarity, in that the category label adds to the similarity of the same-category items. However, this interpretation cannot explain why children use the label only when they believe it accurately reflects category membership (see also Gelman and Waxman 2007, for discussion). If the label refers to a temporary attribute (e.g., "sleepy" rather than "bird"), then children fail to use it (Gelman and Coley 1990).

Likewise, the speaker's communicative intent affects children's willingness to use the label (Jaswal 2004). These findings suggest that the label is a cue to category membership, rather than a perceptual feature per se.

Innateness. A key essentialist assumption is that characteristic features of a category (e.g., a pig's curly tail, or a dog's barking) are innate: they are fixed at birth and unaffected by environmental influences. This notion has been tested with a "switched-at-birth" task. Children learn about a person or animal that has a set of biological parents and then is switched at birth to a new environment and a new set of adoptive parents. Children are then asked to decide whether the offspring will have the features of the birth parents or of the adoptive parents. For example, if a newborn kangaroo is raised by goats, will the animal eventually be good at hopping (the innate choice) or climbing (the upbringing choice)? Preschool children typically report that it would be good at hopping and have a pouch. Although it cannot hop at birth (because it is too young), and it is raised by goats that cannot hop, and it never sees another kangaroo, hopping is inherent to kangaroos; this property will eventually be expressed. Although there is debate as to when precisely this understanding emerges, even on a conservative estimate it appears by about six years of age, and in some studies this understanding appears as early as age four. Innateness beliefs have been documented in children's reasoning about animal categories, plant categories, race categories, gender categories, and (to lesser degree) human traits. One interesting finding is that children tend to be even more nativist than adults (Taylor 1996; Taylor et al. 2008; Hirschfeld and Gelman 1997; Heyman and Gelman 2000a.

Underlying reality. Finally, there is evidence that preschool children expect category members to have internal parts that relate to or cause outward behaviors. When asked to describe the insides of animals, children expect a different set of properties than when asked to describe the insides of inanimate objects (Gelman 1990; Simons and Keil 1995). This is so even when children are asked to consider novel items they have never seen before, and even when the cues to category membership are subtle (for example, animals with no face, such as sea slugs, and artifacts with a face, such as Furbies; Gelman 2003). Furthermore, children are especially likely to appeal to internal parts when they discover that the items unexpectedly turn out to be capable of self-generated movement. This finding suggests that children see the insides of items as having causal force. Preschool children rely more on non-visible causal features than outward appearances to determine category identity (Gopnik and Sobel 2000; Sobel et al. 2007).

From these and other studies, I conclude that children are "essentialists." They believe that certain categories are immutable, allow novel predictions, have innate properties, and have an underlying reality. Essentialism, therefore, does not require specialized scientific knowledge, deep philosophical training, or knowledge of historical contexts.

Where does childhood essentialism come from? The available evidence suggests that essentialism is a basic human bias. It is not explicitly taught by parents or other adults. When provided with an opportunity to explain or discuss atypical categories instances to their young children, parents rarely mention innateness, non-obvious properties, immutability, or defining features (Gelman et al. 1998, 2004). Moreover, essentialism is not limited to Western, industrialized societies. Essentialist beliefs of the sort sketched out above are found in a broad range of cultural contexts, including Madagascar (Astuti et al. 2004), Yucatec Mayan in Mexico (Atran et al. 2001), the Menominee, a Native American group in the U.S. (Waxman et al. 2007), and varying socioeconomic groups in Brazil (Diesendruck 2001; Sousa et al. 2002). Altogether, these results suggest that essentialism emerges in varying environmental contexts, without explicit prompting or instruction.

However, language may also (implicitly) encourage an essentialist stance. All languages have two expressive devices that may be important mechanisms for transmitting essentialism (Gelman 2003): 1) common noun labels (e.g., "this is a bird") express membership in a kind, allowing children to overcome misleading perceptual features and to distinguish between what something is and what something is like; and 2) generic noun phrases (e.g., "Bats live in caves") refer to kinds directly (as extending beyond individual instances) and imply that a feature is relatively enduring and central. Both labeling and generics are ubiquitous in child-directed speech and are acquired early by children (Gelman 2004, 2008; Waxman 2003; Xu 2002). These linguistic devices are insufficient by themselves to transmit essentialism to children. However, they could be important clues that encourage and emphasize an essentialist perspective about particular categories to children who are already capable of reasoning in this way.

2 Conclusions

The research on childhood essentialism suggests that children are not concrete thinkers, focused only on outward appearances. On the positive side, they expect the natural world to have a deeper, hidden reality, which may pave the way for appreciating (and engaging in) science. On the negative side, this characteristic may pave the way for stereotyping social categories and biological misconceptions, as noted earlier.

The study of childhood essentialism also speaks to ongoing debates about the nature of early cognition. There is a rich array of research findings over the past 20 years demonstrating that infants and young children are highly skilled in reasoning about a range of non-obvious features, including number, causality, intentions, goals, and physical principles, such as containment and support (Gelman and Kalish 2006; Cohen and Cashon 2006). Young children's capacity to look beyond the obvious when reasoning about biological categories, as well, is consistent with this research tradition.

Childhood essentialism poses a challenge to the competing (and enduringly popular) view of children as perceptually bound, concrete, and focused on obvious features of their environment. A classic demonstration can be found with Piaget's "conservation" task, in which below roughly six or seven years of age, children believe that pouring a cup of liquid from one container into a taller, thinner container will increase its amount. Children err by focusing on the obvious perceptual feature of the height of the liquid in the container. More recently, a number of theorists have likewise claimed that children rely exclusively on low-level perceptual similarity to classify experience (Smith et al.)1996; Sloutsky and Fisher 2004).

Although differing underlying assumptions in part account for the discrepancy between the views of children as essentialist versus perceptually based (e.g., Gelman and Waxman 2007), another important point is that children use different kinds of information, depending on the concept and the task under consideration. For example, children have more difficulty creating a category than reasoning about a category that has already been established (Gelman 2003). Likewise, tasks that involve non-basic level categories and non-natural kinds, and/or provide no opportunity for children to reason about causation, origins, or innate potential, are unlikely to reveal essentialist effects. In other words, children display different kinds of knowledge depending on the questions they are asked. If we ask preschool children to apply a deterministic rule to a novel array of items for which multiple sorts of classifications are possible, then they will look inconsistent and focused on perceptual features. But if we ask preschool children what it means to be a member of a lexicalized natural kind ("girl," "dog"), a wealth of essentialist assumptions emerges.

Even for adults, essentialist tendencies may persist. Consider once again people's misconceptions about evolution. There are at least four ways that essentialism may pose obstacles to understanding Darwin's ideas. First, the assumption that categories are stable and unable to change is in conflict with the basic principle that species evolve over time (Mayr 1991). Second, the tendency to intensify category boundaries makes it difficult to grasp that two different species may have a common ancestor. This tendency is particularly a difficulty for reasoning about human evolution, as people resist the notion that humans are simply one animal among many (Johnson et al. 1992). Third, essentialism may encourage people to underestimate variation within a category, thereby leading to difficulty with the principle of natural selection (which is deeply rooted in the notion of within-category variation). Fourth, essentialism assumes that causes are inherent in individual organisms (e.g., the essence within each animal), rather than populations. This tendency to focus on inherent properties within an individual rather than characteristics of a population leads to misconstruals of evolution (Sober 1994; Shtulman 2006; Shtulman and Schulz 2008).

As these examples illustrate, essentialist reasoning has far-reaching implications for how humans construe the natural world. One important question for the future is how we can overcome the biases and distortions that essentialism imposes.

Acknowledgments Preparation of this paper was supported by NICHD grant HD36043 and a James McKeen Cattell Fellowship. Correspondence should be addressed to: Susan Gelman, Department of Psychology, University of Michigan, 530 Church St., Ann Arbor MI 48109-1043, e-mail: gelman@umich.edu.

References

Ahn W, Kalish C, Gelman SA, Medin DL, Luhmann C, Atran C, Coley JD, Shafto P (2001) Why essences are essential in the psychology of concepts. Cognition 82:59–69

Ahn W, Flanagan EH, Marsh JK, Sanislow CA (2006) Beliefs about essences and the reality of mental disorders. Psychol Sci 17:759–766

Astuti R, Solomon GEA, Carey S (2004) Constraints on conceptual development. Monographs of the Society for Research in Child Development. Vol. 69(3)

Atran S (1990) Cognitive foundations of natural history. Cambridge University Press, New York

Atran S (1998) Folk biology and the anthropology of science: Cognitive universals and cultural particulars. Behav Brain Sci 21:547–609

Atran S, Medin D, Lynch E, Vapnarsky V, Ek' EU, Sousa P (2001) Folkbiology doesn't come from folkpsychology: Evidence from Yukatek Maya in cross-cultural perspective J Cogn Culture 1:3–42

Berlin B (1992) Ethnobiological classification: principles of categorization of plants and animals in traditional societies. Princeton University Press, Princeton

Bloom P (2000) How children learn the meanings of words. MIT Press, Cambridge

Cohen LB, Cashon CH (2006) Infant cognition. In: Kuhn D, Siegler RS (eds) Handbook of child psychology. Vol 2. Cognition, perception, and language (6th ed.). Wiley, Hoboken, pp 214–251

Coley JD, Medin DL, Proffitt JB, Lynch E, Atran S (1999) Inductive reasoning in folkbiological thought. In: Medin DL, Atran S (eds) Folkbiology. MIT Press, Cambridge, pp 205–232

Diesendruck G (2001) Essentialism in Brazilian children's extensions of animal names. Dev Psychol 37:49–60

Dupré J (1993) The disorder of things: Metaphysical foundations of the disunity of science. Harvard, Cambridge

Evans EM (2001) Cognitive and contextual factors in the emergence of diverse belief systems: creation versus evolution. Cogn Psychol 42:217–266

Fodor J (1998) Concepts: Where cognitive science went wrong. Oxford, New York

Fuss D (1989) Essentially speaking: Feminism, nature, and difference. Routledge, New York

Gelman R (1990) First principles organize attention to and learning about relevant data: Number and the animate-inanimate distinction as examples. Cogn Sci 14:79–106
Gelman SA (2003) The essential child: Origins of essentialism in everyday thought. Oxford, New York
Gelman SA (2004) Learning words for kinds: Generic noun phrases in acquisition. In: Hall DG, Waxman SR (eds) Weaving a lexicon. MIT Press, Cambridge, pp 445–484
Gelman SA, Markman EM (1986) Categories and induction in young children. Cognition 23: 183–209
Gelman SA, Coley JD (1990) The importance of knowing a dodo is a bird: Categories and inferences in 2-year-old children. Dev Psychol 26, 796–804
Gelman SA, Wellman HM (1991) Insides and essences: Early understandings of the nonobvious. Cognition 38:213–244
Gelman SA, Heyman GD (1999) Carrot-eaters and creature-believers: the effects of lexicalization on children's inferences about social categories. Psychol Sci10:489–493
Gelman SA, Hirschfeld LA (1999) How biological is essentialism? In: Atran S, Medin D (eds) Folkbiology. MIT Press, Cambridge
Gelman SA, Kalish CW (2006) Conceptual development. In: Kuhn D, Siegler R (eds) Handbook of child psychology. Vol. 2: Cognition, perception and language. Wiley, New York, pp 687–733
Gelman SA, Waxman SR (2007) Looking beyond looks: comments on Sloutsky, Kloos, and Fisher (2007). Psychol Sci 18:554–555
Gelman SA, Coley JD, Rosengren K, Hartman E, Pappas A (1998) Beyond labeling: The role of maternal input in the acquisition of richly-structured categories. Monographs of the Society for Research in Child Development. Serial No. 253, Vol. 63, No. 1
Gelman SA, Taylor MG, Nguyen SP (2004) Mother-child conversations about gender. Monographs of the Society for Research in Child Development. Vol 69(1)
Gelman SA, Goetz PJ, Sarnecka BS, Flukes J (2008) Generic language in parent-child conversations. Lang Learn Dev, in press
Gil-White FJ (2001) Are ethnic groups biological "species" to the human brain? Curr Anthropol 42:515–554
Gopnik A, Sobel DM (2000) Detecting blickets: How young children use information about novel causal powers in categorization and induction. Child Dev 71:1205–1222
Graham SA, Kilbreath CS, Welder AN (2004) Thirteen-month-olds rely on shared labels and shape similarity for inductive inferences. Child Dev 75:409–427
Guillaumin C (1980) The idea of race and its elevation to autonomous scientific and legal status. In Sociological theories: race and colonialism. UNESCO, Paris, pp 37–68
Haslam N, Ernst D (2002) Essentialist beliefs about mental disorders. J Soc Clin Psychol 21:628–644
Haslam N, Rothschild L, Ernst D (2000) Essentialist beliefs about social categories. Brit J Soc Psychol 39:113–127
Haslam N, Rothschild L, Ernst D (2002) Are essentialist beliefs associated with prejudice? Brit J Soc Psychol 41:87–100
Heyman GD, Gelman SA (2000a) Beliefs about the origins of human psychological traits. Dev Psychol 36:665–678
Heyman GD, Gelman SA (2000b) Preschool children's use of traits labels to make inductive inferences. J Exp Child Psychol 77:1–19
Hirschfeld LA, Gelman SA (1997) What young children think about the relation between language variation and social difference. Cogn Dev 12:213–238
Inspector Y, Kutz I, David D (2004) Another person's heart: Magical and rational thinking in the psychological adaptation to heart transplantation. J Psych Related Sci 41:161–173
Jaswal VK (2004) Don't believe everything you hear: Preschoolers' sensitivity to speaker intent in category induction. Child Dev 75:1871–1885
Jaswal VK, Markman EM (2007) Looks aren't everything: 24-month-olds' willingness to accept unexpected labels. J Cogn Dev 8:93–111

Jayaratne T (2001) National sample of adults' beliefs about genetic bases to race and gender. Unpublished raw data
Johnson K, Mervis C, Boster J (1992) Developmental changes within the structure of the mammal domain. Dev Psychol 28:74–83
Keil F (1989) Concepts, kinds, and cognitive development. MIT Press, Cambridge
Mayr E (1991) One long argument: Charles Darwin and the genesis of modern evolutionary thought. Harvard University Press, Cambridge
Medin DL, Ortony A (1989) Psychological essentialism. In: S Vosniadou, A Ortony (eds) Similarity and analogical reasoning. Cambridge, New York, pp 179–195
Mervis CB, Rosch E (1981) Categorization of natural objects. Ann Rev Psychol 32:89–115
Pinker S (1994) The language iinstinct. W. Morrow, New York
Prasada S (2000) Acquiring generic knowledge. Trends Cogn Sci 4:66–72
Prentice DA, Miller DT (2007) Psychological essentialism of human categories. Curr Directions Psychol Sci 16:202–206
Rehder B (2007) Essentialism as a generative theory of classification. In: Gopnik A, Schulz L (eds) Causal learning: psychology, philosophy, and computation. Oxford University Press, New York, pp 190–207
Schwartz SP (ed) (1977) Naming, necessity, and natural kinds. Cornell University Press, Ithaca
Shtulman A (2006) Qualitative differences between naïve and scientific theories of evolution. Cogn Psychol 52:170–194
Shtulman A, Schulz L (2008) The relationship between essentialist beliefs and evolutionary reasoning. Cogn Sci, in press
Simons DJ, Keil FC (1995) An abstract to concrete shift in the development of biological thought: the insides story. Cognition 56:129–163
Sloutsky VM, Fisher AV (2004) Induction and categorization in young children: A similarity-based model. J Exp PsycholGeneral 133:166–188
Smith EE (1989) Concepts and induction. In: Posner MI (ed) Foundations of cognitive science. MIT Press, Cambridge, pp 501–526
Smith LB, Jones SS, Landau B (1996) Naming in young children: A dumb attentional mechanism? Cognition 60:143–171
Sobel DM, Yoachim CM, Gopnik A, Meltzoff AN, Blumenthal EJ (2007) The blicket within: preschoolers' inferences about insides and causes. J Cogn Dev 8:159–182
Sober E (1994) From a biological point of view. Cambridge, New York
Sousa P, Atran S, Medin D (2002) Essentialism and folkbiology: Evidence from Brazil. J Cogn Culture 2:195–223
Strevens M (2000) The essentialist aspect of naive theories. Cognition 74:149–175
Sylvia C, Novak W (1997) A change of heart. Little, Brown, Boston
Taylor MG (1996) The development of children's beliefs about social and biological aspects of gender differences. Child Dev 67:1555–1571
Taylor MG, Rhodes M, Gelman SA (2008) Boys will be boys; cows will be cows: Children's essentialist reasoning about gender categories and animal species. Unpublished ms.
Waxman SR (2003) Links between object categorization and naming: origins and emergence in human infants. In: Rakison DH, Oakes LM (eds) Early category and concept development: making sense of the blooming, buzzing confusion. Oxford, New York
Waxman S, Medin DR, Ross N (2007) Folkbiological reasoning from a cross-cultural developmental perspective: early essentialist notions are shaped by cultural beliefs. Dev Psychol 43:294–308
Wilson RA (1999) Realism, essence, and kind: Resuscitating species essentialism? In: Wilson RA (ed) Species: new interdisciplinary essays. MIT Press, Cambridge, pp 187–207
Xu F (2002) The role of language in acquiring object kind concepts in infancy. Cognition 85:223–250

The Human Brain "Projects" upon the World, Simplifying Principles and Rules for Perception

Alain Berthoz

Abstract "Each subject lives in a world where there is only of subjective realities and where the same environments represent only subjective realities." Von Uexküll.[1]

1 Introduction: The Concept of Umwelt

"All the features of objects are in fact nothing else than the perceptive characters which are attributed to them by the subject with whom they have a relation." This statement by von Uexküll[2] summarizes one of his essential ideas, but an even more interesting and modern thought was what he called the "lived world," which translates into French as "monde vécu" by a subject who acts in this world. This reversal of the classical description of the mechanisms of perception and action places the intentional and goal-oriented subject at the origin of the process. The subject builds his world according to his basic needs and action tools. This view has also been promoted by Bergson and Husserl. I have also proposed that the "projective brain" is a simulator and an emulator of reality that builds its perceived world according to its planned acts and also that this new view is essential for understanding intersubjectivity (Berthoz 1997; Berthoz and Petit 2003).

The human brain imposes, in a top-down fashion, its rules of interpretation on sensory data. It transforms the perceived world according to rules of symmetry, stability, and kinematic laws derived from principles of maximum smoothness. These rules follow simplifying principles that allow the simplification of neurocomputation to speed up action. Top-down controlled attention is also a powerful selector.

A. Berthoz
Collège de France/CNRS Paris,
11 Place Marcelin Berthelot, 75005 Paris, France
e-mail: Alain.berthoz@college-de-France.fr

[1] J.V. Uexküll. Mondes animaux et monde humain. Denoël 1965. (The English translation is mine). p. 85.

[2] J.V. Uexküll.op. cit.

A. Berthoz and Y. Christen (eds.), *Neurobiology of "Umwelt": How Living Beings Perceive the World,* Research and Perspectives in Neurosciences,

The challenge for neuroscience today is to provide evidence concerning these speculations. I will in this presentation present some empirical data that suggest that the concept of Umwelt is applicable to man and that we are also, in a sense, prisoners of our "monde vécu." More generally, as hypothesized by von Uexküll, the intimate resonance between action and perception is not only proven by the recent discovery of mirror neurons but also the fact that the laws of physics, like Newton's laws, are implemented in the brain and serve as anticipatory mechanisms of action. In addition, specialized modules analyze specific properties of the environment that correspond to specific tasks such as the need to recognize the face and its emotional expressions, shape, or movement of living creatures.

Strictly speaking "Um-welt" means the "world around" in which animals and humans live. It can be translated in French by "Milieu." However, for von Uexküll it includes the world of things in the environment, the perceived world, thc signals emitted by both the subject and the things, and the actions that can be performed by each species. Above all, it includes the *significance* or meaning of things for each animal, in that they are potentially participating in the survival and social relations of the animal.

It is therefore a dynamic, interactive concept defining the relations between the physical world and living organisms, leading to *intersubjectivity*. For von Uexküll a living organism is not a machine (e.g., a watch). It is not built in a "centripetal" way by assembling parts but, on the contrary, starts from a principle, a goal, or an intentionthat, in a "centrifugal" way, determines the fitting of the parts; hence the idea that the brains of animals "project" properties on the world. They predict and anticipate. The brain does not passively receive visual information; it decides to solve ambiguities.

The brain imposes symmetry on the visual world (a beautiful example was given by psychologists in California who designed the famous Ames chambers, showing that the brain imposes symmetry on the visual world). There are many other examples of how the human brain imposes rules of interpretation upon the world and, as I proposed in my book, of the brain's sense of "movement:" illusions are solutions in the cases where the perceived world is ambiguous. We could say in the frame of this analysis that the chosen solutions are dependent upon the Umwelt of each of us.

This view has been recently updated by theories about the way the brain predicts (Changeux 1983; Llinas 2001). Von Uexküll has also placed his theory in the perspective of evolution. He thinks that the selection of relevant information made by each species is aimed at the survival of the normal individual in the interest of the prolongation of the species itself.

2 Action Specifies the Selection of Meaningful Perception

One of the fundamental ideas contained in the concept of Umwelt is that each animal simplifies the perception of the outside world, selecting relevant information that registers with its repertoire of possible actions. In other words, the action repertoire of each species specifies the selection of meaningful, significant information that is being searched for actively. A "world" is therefore created that is made of those aspects of the environment that are important for the behavioral repertoire, or survival tools, given to each animal by evolution. Simplification is induced by this selection.

Von Uexküll writes that the bees perceive only two visual forms: open (which attracts them) or closed (which induces a flight behavior). The jackdaw does not perceive the motionless grasshopper[3]."The ground worms do not guide their action according to the form but according to the taste of the leaf. Because the sensors of the ground worm have a structure too simple to work out the perceptive characters of the form." The senses provide only one very simplified version of the objects or interesting creatures. "Only the phenomena which have a meaning for the animal subject are changed into nervous excitations."[4] (p. 124). "It is the meaning which is the leading concept with which biology must be guided, and not the poor wretched rule of causality which cannot further see more than a step ahead or a step behind, and remains blind with respect to great structural relationships ""I claim that the original partition of the fly (which one can also call his original image), its prototype, acts in such a way on the original partition of the spider that the web of the spider can be described as "flycatcher."[5]

For von Uexküll, "Each action,..., prints its signification on any neutral object and transforms it, in each world into a carrier of significance".[6] "The stem of a flower can be transformed in four different by specified environments by respectively a young girl, the ant, the larva, the cicada and the cow. It is, for each one, carrying significance because they respectively use the stem like an ornament, a road to reach the leaf, a construction material, and a food." The "form" is therefore the product of the acts of a subject. He writes: "The meaningful form, that which lasts, is always the product of a subject: it is never the product of a subject subjected to a anomic action, whatever is the duration of this action."[7]

This interpretation is extremely interesting in light of modern Neuroscience because it links very closely perception and action and actually places action at the origin of perception. The discovery of mirror neurons (Rizzolatti and Craighero 2004), which code both the action of the subject and the action of others, also calls for a repertoire of perceptions linked with the repertoire of actions of each individual. If we follow this reasoning, it could be speculated that each species may have a different repertoire of mirror neurons.

What can be the advantage of such a reduction and selection of perception? Firstly, to ensure the relevant action necessary to survival, for instance, the fact that the tick recognizes preferably Butyric acid and warm skin allows it to find the blood it needs for feeding. Secondly, it simplifies neurocomputation and allows rapid reaction to environmental hazards, prey catching or escape from predators. For

---

[3] Von Uexküll op.cit. p. 73.

[4] Von Uexküll op. cit. P. 124.

[5] Von Uexküll, op.cit. p. 116.

[6] Von Uexküll op. cit. p. 100.

[7] Von Uexküll op. cit. p. 107.

example, butterflies can only detect the sounds emitted by their enemies and the rest of the world is silent; for instance, they only hear the cry of bats, which allows them to identify their predators in the complex environment that surrounds them. But it also has a more sophisticated advantage: the general understanding of the role of each object of the world in the action plan of the subject.

A well-known example given by Von Uexküll is that of the young black man who had never seen a ladder and who at first perceived it only as "sticks" and "holes." However, when he saw somebody going up, this object was suddenly perceived of as a "ladder." This example showd that we perceive objects starting not from a "mental image" but from a "potential action" done with the object (a motor schema). Von Uexküll writes: "The significance is conferred by the act of the subject"[8] (p.95). Similar ideas can be found in the phenomenological tradition in the writings of Husserl, Merleau-Ponty and in the concept of "enaction" of Varela (Varela et al. 1991). J.L. Petit and I have reviewed these ideas in our book on the relation between phenomenology and physiology of action (Berthoz and Petit 2003).

From these ideas stems the following proposal: the lived body (corps vécu) is fundamental in creating an Umwelt, even in humans. And there is no perception of the world that does not refer somehow to the acting body. This theory requires very elaborate systems of reference frames. The common house of the Mirana, the "maloca" (Karadimas 2006) is built following the architecture of two sexually defined bodies: the sky is male, the earth is female. Three spaces constitute the Umwelt of the Maloca house: the geographical space, the house, and the sexual body.

In other words, what I would like to suggest is that the selection of meaningful percepts is not only a problem of selecting sensors and stimuli. Because it is always related to some action, it also requires very sophisticated selection and manipulation of spatial reference frames. Von Uexküll had the intuition that, to simplify the problem of reference frames, it was necessary for the brain to have available a basic reference frame. He attributed to the vestibular system a fundamental role in establishing a common reference frame for perception and action.

3 The Problem of Reference Frames

For von Uexküll there is a fundamental reference frame: the vestibular system. He writes: "All the animals which have three circular semi canals also have a notion of three-dimensional space"[9] This statement was also made by Poincaré, who wrote that Japanese mice, which have only two semi-circular canals, also only have a perception of two-dimensional space!

These ideas of Poincaré and von Uexküll were very similar in that they thought that space was an active construct: "Active space is not only a space of movement made up from thousand of steps in all directions but it has a system

[8] Von Uexküll op. cit. p. 95.

[9] Von Uexküll op. cit. p. 31 and 32.

of coordinates, which is used as a basis for all the space determinations. It is of capital importance that whoever deals with the problem of space is convinced of this fact. Any normal human carries with him a frame of reference formed of three planes of the three semi-circular canals."[10] But for von Uexküll this system also allows "homing" (path integration) and more generally the memory of traveled paths.

The intuition of von Uexküll concerning the role of the vestibular system was right: it plays a fundamental role in higher cognitive processes in relation to the environment. He had an intuition about what has been recently discovered, i.e., the vestibular contribution to path integration, and more generally, to navigation. Deficits in the construction of a coherent self to which the vestibular system contributes may play a role in the symptoms of agoraphobia and spatial anxiety, autism, schizophrenia and a number of psychiatric diseases involving relations with others (empathy).

Today we know that the vestibular system is not only involved in reflex actions - such as the vestibulo-ocular reflex, which stabilizes the image of thc world on the retina - or vestibule-spinal reflexes, which allow the stabilization of posture. Vestibular signals are carried through the vestibular nuclei to the sensory thalamus and from there are sent to an area called the vestibular cortex, which has been identified by fMRI studies (Bottini et al. 1994; Lobel et al. 1998; Brandt et al. 1980) to be located at the temporo-parietal junction, although other groups have located it in the posterior insula. It is now accepted that the vestibular system plays a fundamental role in spatial orientation; it is also involved in the memory of traveled paths (Berthoz et al. 1989, 1995; Israel et al. 1993, 1997a,b; Israel and Berthoz 1989; Nico et al. 2002).

However, the vestibular system is also fundamental in the elaboration of what Penfield characterized as "body awareness and spatial relationship"(Penfield 1957). Recent data from our laboratory (Kahane et al. 2003) and from other groups have shown that the stimulation of the temporo-parietal junction induced "out of body experiences," and the stimulation of areas along the peri-sylvian cortex induced a variety of illusions of rotations or elevations. It is therefore now evident that the vestibular cortex plays a fundamental role in many aspects of our relations with the outside world, and vestibular asymmetries or anomalies induce a number of perceptual deficits (Berthoz and Rousie 2001; Rousie et al. 1999). In addition the vestibular cortex seems to be involved in the elaboration of an "internal model" of Newton's law (Indovina et al. 2005; McIntyre et al. 2001; Zago and Lacquaniti 2005). It seems that, during childhood, we build a capacity to internally simulate gravity and its influence upon the movement of objects. Phrased in the terms of von Uexküll, we simulate internally a very important property of the physical world around us, which is necessary to perform our actions against the forces of gravity.

[10] Von Uexküll op.cit. p. 31.

4 Simplifying Principles Decreasing "Neurocomputation" Time and Complexity

I would like to propose the idea, implicitly contained in von Uexküll descriptions, that one of the essential reasons for the restriction of each species' Umwelt is due not only to the matching of each species' need and action repertoires but also to the need to reduce neurocomputation to achieve two main goals: speed and robustness. This, I think, is obtained through the selection, in the course of evolution, of simplifying principles that optimize the perception-action process and minimize or even suppress the "computation" needed. The simplifying principles and mechanisms used by living organisms are numerous and I will only mention a few here. The price for these simplifications is, of course, the reduction of the understanding we have about the world; it creates an Umwelt. Let me give a few examples.

Bernstein proposed that simplification was obtained for the control of movement by mechanisms of reduction of the degrees of freedom. Gibson and his followers also proposed this idea from the "ecological" theories and the concept of affordance. For example, to climb a stair the brain does not measure distances but *relations* between body size and obstacle size. The brain uses relevant variables that give it instant access to important parameters. For example, to detect moving objects the brain does not estimate distance but "time to contact" (Lee 1976). Another extremely interesting mechanism is the rapid detection of fearful shapes by the amygdala (Armony and LeDoux 1997; LaBar et al. 1998; Morris et al. 1998). The need to recognize very rapidly natural forms has also probably prompted the design of specialized areas of the brain for recognition of living creatures. Fize et al.2000 have shown that a part of the fusiform area is specialized in the rapid detection of animal faces; other areas are specialized in the detection of the body (the extrastriate body area, for instance) or even environmental forms (the parahippocampus, for instance). Although this hypothesis has not been explored so far, it can be speculated that each species has a repertoire of canonical forms that it can recognize rapidly according to the meaning of that form for each species in term of prey or predators. This ability contributes to the specification of the Umwelt of each species. Early learning during infancy probably also influences this repertoire, as shown by the ethologists for the phenomenon of "imprinting." The imprinting on the shape of the mother certainly simplifies the interaction of the baby animal with the environment.

Another important aspect of Umwelt is the nature of visual space. It is well known that each species has a different analysis of visual space and that visual perception of the world influences behavior in primates (Cheney and Seyfarth 1990). Humans may not use a Euclidian visual space; they may process visual information using affine geometry (Koenderink and van Doorn 1991) or even more complex geometries (Bijl and Koenderink 1993; Kappers et al. 1994; Koenderink and van Doorn 1991, 2000; Koenderink et al. 2002). These geometries seem complex but in fact they simplify neurocomputation

Simplification is also performed centrally by a principle of specialization of processes. A nice example is given by the fact that, in the brain, a very fast pathway identifies natural forms of animals or human forms like faces (in the fusiform area; Baker et al. 2007; Kanwisher and Yovel 2006), bodies (in the extrastriate body area; Arzy et al. 2006; Lamm and Decety 2008; Urgesi et al. 2007), environmental forms (in the parahippocmapus; Epstein et al. 1999), etc. Another important aspect of the brain's specialization that participates in the construction of our Umwelt is the mirror system (Rizzolatti 2005). Mirror neurons respond to specific gestures made either by the subject or by others, indicating that the brain contains a repertoire of predetermined gestures that are, as von Uexküll predicted, meaningful for each species.

5 Vicariousness

Von Uexküll insists upon the fact that living organisms often carry out the same action in many very different ways. (He describes the famous example of the ursin.

This flexibility of action patterns and cognitive strategies is also found in humans. Von Uexküll wrote, "The world in which an animal lives, and that we see extending around him, changes, when one looks from the point of view of the animal itself, in its environment, in the space where are acting the various carriers of meaning"[14] Humans have the ability to change points of view and escape from an egocentric view of the world.

An interesting example of this flexibility is the mechanisms of spatial memory for navigation. Mice or rats can use egocentic or allocentric strategies for navigation (Trullier et al. 1997; Burguiere et al. 2005). The human brain has a repertoire of cognitive strategies that is even more flexible (Amorim et al. 1997; Committeri et al. 2004; Galati et al. 2000; Lambrey and Berthoz 2007; Mellet et al. 2000; Schmidt et al. 2007; Vidal et al. 2004). The activity in parahippocampus and retrosplenial cortex is related to a changing point of view. The right hippocampus has become mainly specialized in allocentric coding of spatial relations and events. The left hippocampus has become specialized in sequential egocentric memory of travelled routes and episodic memory (which fits with the language description of itineraries), and this may correspond to hemispheric lateralization for spatial competences (global on the right; detailed, categorical on the left), although this point is still debated.

6 Gender Differences: are the Umwelts of Men and Women Different?

"The striking quantity and diversity of sex-related influences on brain function indicate that the still widespread assumption that sex influences are negligible cannot be justified, and probably retards progress in our field" (Cahill 2006). Gender differences are another aspect of the human, and possibly animal, Umwelt. Although this question is largely debated, it seems that men and women really do not apprehend the world in the same way. There are anatomical and neuroendocrinal bases for gender differences that are out of the scope of this paper, but it is now accepted that, whatever the origin (nature or nurture) of these differences, men and women do not process spatial information, for instance, in the same way. These gender differences are probably related, in a way that is still obscure, to the fact that large gender differences are observed in the impact of psychiatric perturbations (spatial anxiety, agoraphobia etc.) more frequently among women. For virtual navigation, women have been found to use a parieto-frontal system (ego); men use a parieto-hippocampal system. But there is no difference in navigating an eight-arm maze! Men perform better on a transfer from virtual to real space. Gender differences are also found in the use of external landmarks versus spatial representations updated by self-motion (Lambrey and Berthoz 2007). It has been suggested that spatial mental representations of large-scale environments contain more metric information in men than in women but contain more landmark information in women than in men. We found that men readily relied on an internal, egocentric representation of where landmarks were expected to be perform a pointing task, a representation that could be updated by a path integration-like process during self-motion (spatial updating). In contrast, women seemed to take their bearings more readily on the basis of the presumed stable landmarks of the external world. We suggest that this difference in spatial orientation strategy may explain why environmental representations contain more metric information in men than in women, since spatial updating necessarily requires the use of metric movement information, namely angles and distances relative to self-motion.

An interesting consequence of the differences in spatial Umwvelt is that it may also have an impact on social relations and the way we perceive others. I have proposed a theory of empathy (Berthoz 2004) that includes an important role for the capacity of the human brain to change spatial perspective.

7 The "Magic Umvelts"

Von Uexküll is interested in the fact that animals can create imaginary worlds (sometimes innate) that allow predictive behavior. He calls these "magic Umwelts:" "They are umwelts created by the subject (we would undoubtedly say today imaginary worlds or hallucinations or virtual worlds (as the dream)....that the subject itself is alone to perceive" [11]. He gives an example: the traveled path. "In the familiar path a series of perceptive signals established from memory by former experiences, take turns mutually, whereas in the innate way the same series of signals is immediately given as magic appearance...." We would probably not call this magic today, but the challenge of modern neuroscience, and of this meeting, is precisely to understand the biological basis of our Umvelt and the abilities we have to escape the innate limitations of our dialogue with the world.

[11] Von Uexküll op. cit. p. 81.

References

Amorim MA, Glasauer S, Corpinot K, Berthoz A (1997) Updating an object's orientation and location during non visual navigation: a comparison between two processing modes. Percept Psychophys 59:404–418

Armony JL, LeDoux JE (1997) How the brain processes emotional information. Ann NY Acad Sci 821:259–270

Arzy S, Thut G, Mohr C, Michel CM, Blanke O (2006) Neural basis of embodiment: distinct contributions of temporoparietal junction and extrastriate body area. J Neurosci 26:8074–8081

Baker CI, Hutchison TL, Kanwisher N (2007) Does the fusiform face area contain subregions highly selective for nonfaces? Nature Neurosci 10:3–4

A. Berthoz,. Le sens du mouvement O. Jacob 1997 (English translation: The brain's sense of movement. Harvard Univ Press 2000)

Berthoz A (2004) L'empathie. Paris O. Jacob

A. Berthoz La décision. O.Jacob 2003 (English translation: Emotion and reason: the cognitive foundations of decision making. Oxford Univ. Press. 2006)

A. Berthoz, J.L. Petit. Phénoménologie et Physiologie de l'action. O. Jacob. 2006 (English translation : Oxford Univ. Press. 2008)

Berthoz A, Israel I, Zee DS, Vitte E (1989) Linear displacement can be derived from otholic information and stored on spatial maps controlling the saccadic system. Adv Oto-Rhino-Laryngol 41:76–81

Berthoz A, Israel I, Georges-Francois P, Grasso R, Tsuzuku T (1995) Spatial memory of body linear displacement: what is being stored? Science 269:95–98

Berthoz A, Rousie D (2001) Physiopathology of otolith-dependent vertigo. Contribution of the cerebral cortex and consequences of cranio-facial asymmetries. Adv Otorhinolaryngol 58:48–67

Bijl P, Koenderink JJ (1993) Visibility of elliptical Gaussian blobs. Vision Res 33:243–255

Bottini G, Sterzi R, Paulelscu E, Vallar G, Cappa S, Erminio F, Passingham RE, Frith CD, Frackowiack RSJ (1994) Identification of the central vestibular projections in man: a positron emission tomography activation study. Exp Brain Res 99:164–169

Brandt T, Arnold F, Bles W, Kapteyn TS (1980) The mechanism of physiological height vertigo. I. Theoretical approach and psychophysics. Acta Otolaryngol (Stockh) 89:513–523

Burguiere E, Arleo A, Hojjati M, Elgersma Y, De Zeeuw CI, Berthoz A, Rondi-Reig L (2005) Spatial navigation impairment in mice lacking cerebellar LTD: a motor adaptation deficit? Nature Neurosci 8:1292–1294

Cahill L (2006) Why sex matters for neuroscience. Nature Rev Neurosci 7:477–484

J.P. Changeux, L'homme neuronal. Fayard (1985). R. Llinas. I of the vortex. From neurons to self. 2001. MIT Press

Committeri G, Galati G, Paradis AL, Pizzamiglio L, Berthoz A, LeBihan D (2004) Reference frames for spatial cognition: different brain areas are involved in viewer-, object-, and landmark-centered judgments about object location. J Cogn Neurosci 16:1517–1535

Cheney DL, Seyfarth RM (1990) How monkeys see the world. Chicago, Chicago Univ Press

Epstein R, Harris A, Stanley D, Kanwisher N (1999) The parahippocampal place area: recognition, navigation, or encoding? Neuron 23:115–125

Fize D, Boulanouar K, Chatel Y, Ranjeva JP, Fabre-Thorpe M, Thorpe S (2000) Neuroimage 11:634–43

Galati G, Lobel E, Vallar G, Berthoz A, Pizzamiglio L, LeBihan D (2000) The neural basis of egocentric and allocentric coding of space in humans: a functional magnetic resonance study. Exp Brain Res, Jul;133:156–64

Indovina I, Maffei V, Bosco G, Zago M, Macaluso E, Lacquaniti F (2005) Representation of visual gravitational motion in the human vestibular cortex. Science 308:416–419

Israel I, Berthoz A (1989) Contribution of the otoliths to the calculation of linear displacement. J Neurophysiol 62:247–263

Israel I, Chapuis N, Glasauer S, Charade O, Berthoz A (1993) Estimation of passive horizontal linear whole-body displacement in humans. J Neurophysiol 70:1270–1273

Israel I, Grasso R, Georges-Francois P, Tsuzuku T, Berthoz A (1997a) Spatial memory and path integration studied by self-driven passive linear displacement. I. Basic properties. J Neurophysiol 77:3180–3192

Israel I, Grasso R, Georges-Francois P, Tsuzuku T, Berthoz A (1997b) Spatial memory and path integration studied by self-driven passive linear displacement. I. Basic properties. J Neurophysiol 77:3180–3192

Kahane P, Hoffmann D, Minotti L, Berthoz A (2003) Reappraisal of the human vestibular cortex by cortical electrical stimulation study. Ann Neurol 54:615–624

Kanwisher N, Yovel G (2006) The fusiform face area: a cortical region specialized for the perception of faces. Philos Trans R Soc Lond B Biol Sci 361:2109–2128

Kappers AM, Koenderink JJ, te Pas SF (1994) Haptic discrimination of doubly curved surfaces. Perception 23:1483–1490

D. Karadimas La raison du corps Ed Peters 2008

Koenderink JJ, van Doorn AJ (1991) Affine structure from motion. J Opt Soc Am A 8:377–385

Koenderink JJ, van Doorn AJ (2000) Direct measurement of the curvature of visual space. Perception 29:69–79

Koenderink JJ, van Doorn AJ, Kappers AM, Todd JT (2002) Pappus in optical space. Percept Psychophys 64:380–391

LaBar KS, Gatenby JC, Gore JC, LeDoux JE, Phelps EA (1998) Human amygdala activation during conditioned fear acquisition and extinction: a mixed-trial fMRI study. Neuron 20:937–945

Lambrey S, Berthoz A (2007) Gender differences in the use of external landmarks versus spatial representations updated by self-motion. J Integr Neurosci 6:379–401

Lamm C, Decety J (2008) Is the Extrastriate Body Area (EBA) Sensitive to the Perception of Pain in Others? Cereb Cortex doi:10.1093/cercor/bhn006

Lee DN (1976) A theory of visual control of braking based on information about time-to-collision. Perception 5:437–459

Lobel E, Kleine JF, Bihan DL, Leroy-Willig A, Berthoz A (1998) Functional MRI of galvanic vestibular stimulation. J Neurophysiol 80:2699–2709

McIntyre J, Zago M, Berthoz A, Lacquaniti F (2001) Does the brain model Newton's laws? Nat Neurosci. 2001 Jul;4:693–694

Mellet E, Bricogne S, Tzourio-Mazoyer N, Ghaëm O, Petit L, Zago L, Etard O, Berthoz A, Mazoyer B, Denis M (2000) Neural correlates of topographical mental exploration: the impact of route versus survey perspective learning. Neuroimage 2000;12:588–600

Morris JS, Friston KJ, Buchel C, Frith CD, Young AW, Calder AJ, Dolan RJ (1998) A neuromodulatory role for the human amygdala in processing emotional facial expressions. Brain 121 (Pt 1):47–57

Nico D, Israel I, Berthoz A (2002) Interaction of visual and idiothetic information in a path completion task. Exp Brain Res 146:379–382

Penfield W (1957) Vestibular sensation and the cerebral cortex. Ann Otol (St Louis) 66:691–698

Rizzolatti G (2005) The mirror neuron system and its function in humans. Anat Embryol (Berl) 210:419–421

Rizzolatti G, Craighero L (2004) The mirror-neuron system. Annu Rev Neurosci 27:169–192

Rousié D, Hache JC, Pellerin P, Deroubaix JP, Van TP, Berthoz A (1999) Oculomotor, postural, and perceptual asymmetries associated with a common cause. Craniofacial asymmetries and asymmetries in vestibular organ anatomy. Ann NY Acad Sci 871:439–446

Schmidt D, Krause BJ, Weiss PH, Fink GR, Shah NJ, Amorim MA, Muller HW, Berthoz A (2007) Visuospatial working memory and changes of the point of view in 3D space. Neuroimage 36:955–968

Trullier O, Wiener S, Berthoz A, Meyer JA (1997) Biologically-based artificial navigation systems. Reviews and prospects. Prog Neurobiol 51:483–544

Urgesi C, Candidi M, Ionta S, Aglioti SM (2007) Representation of body identity and body actions in extrastriate body area and ventral premotor cortex. Nature Neurosci 10:30–31

Varela FJ, Thompson E, Rosch E (1991) The embodied Mind: cognitive science and human experience. The MIT Press, Cambridge, USA
Vidal M, Amorim MA, Berthoz A (2004) Navigating in a virtual three-dimensional maze: how do egocentric and allocentric reference frames interact? Brain Res Cogn Brain Res 19:244–258
Zago M, Lacquaniti F (2005) Internal model of gravity for hand interception: parametric adaptation to zero-gravity visual targets on Earth. J Neurophysiol 94:1346–1357

Umwelt: A Psychomotor Functional Event

Rodolfo R. Llinás

Abstract My basic posture concerning Umwelt (world view) is based on the assumption that our perception and understanding of "universals" derives from the functional properties of our brains. Such universals are ultimately constructed by the functional state we know as consciousness. From such a brain-centric perspective, Umwelt is what our brain makes from the sensory inputs arising from their responses to the external world and the ancestral brain network derived from our evolutionary history. Ultimately, then, our Umwelt derives from the sensory specification of internal brain function, mostly determined genetically and epigenetically during development and honed by the leaning process.

1 The Basic Neuronal Circuit in Human Consciousness

Given that sensory inputs generate but a fractured representation of universals, the issue of perceptual unity concerns the mechanisms that allow these different sensory components to be gathered into one global image. In recent years, this gathering has been described as "binding," to be implemented by temporal conjunction (Bienenstock and Von der Malsburg 1986; Von der and Malsburg 1981; Gray et al. 1989; Llinás 1990).

Given that the perceptions category encompasses such an extensive set, a purely hierarchical connectivity matrix, where a small group of neurons represent specific elements of a category (grandmother cell/cluster), is a very unlikely solution to such a problem. A second problem with the hierarchical proposal is sampling size, i.e., a very large number of specific elements in a very large number of categories would render the retrieval problem insurmountable and labile. Thus, even considering that

R.R. Llinás
Department of Physiology and Neuroscience,
New York University Medical School, New York NY 10016 USA
e-mail: llinar01@popmail.med.nyu.edu

A. Berthoz and Y. Christen (eds.), *Neurobiology of "Umwelt": How Living Beings Perceive the World,* Research and Perspectives in Neurosciences,

neuronal elements transduce and transmit signals at a millisecond rate from the onset of sensory primitives, exhausting all sequential combinations would be awkwardly time intensive. However, at a more familiar level, it takes roughly the same amount of time to recognize that a face is familiar than that it is not. As in any sequential strategy, it take much longer to conclude nonfamiliarity, as it would require comparing it with "all known faces," than familiarity, because with the latter the search will proceed for only as long as necessary to match. From a different perspective then, the hypothesized hierarchical connectivity matrix fails to explain how their unique "perceptual insights" (the specific elements in a given category) are communicated to the rest of the nervous system; how do grandmother cells tell the rest of the neurons what they know, given their unique position at the top of a hierarchy?

Alternatively, since categorizations are generated by spatial mapping of the primary sensory cortices and their associated cortical structures, a more dynamic interaction based on temporal coherence may generate dissipative functional structures (Llinás and Paré 1996) capable of as rapid a change as the perception they generate. Thus, a simultaneity mapping may be envisioned that takes advantage of the parallel and synchronous organization of the brain networks to generate perception.

The hypotheses to be discussed below are derived from two areas of research: first, from the investigation of single neuronal elements studied in vitro and in vivo, and second, from measurements made via noninvasive magnetoencephalography in humans. The principal issue to be discussed is the assumption that the intrinsic electrical properties of neurons, and the dynamic events resulting from their connectivity, result in the global resonant states that we know as cognition.

2 Introspection and Reality Emulation

Several lines of research suggest that the brain is essentially a closed system (Llinás and Paré 1996) capable of self-generated activity based on the intrinsic electrical properties of its component neurons and their connectivity. In such a view, the CNS is a "reality"-emulating system (Llinás and Ribary 2001) and the parameters of such "reality" are delineated by the senses (Llinás and Paré 1991). The hypothesis that the brain is a closed system follows from the observation that the thalamic input from the cortex is larger than that from the peripheral sensory system (Wilson et al. 1984), suggesting thalamocortical iterative recurrent activity is the basis for conciseness (Llinás and Paré 1991). In addition, neurons with intrinsic oscillatory capabilities that reside in this complex synaptic network allow the brain to self-generate dynamic oscillatory states that shape the functional events elicited by sensory stimuli. In this context, functional states such as wakefulness or REM sleep and other sleep stages are prominent examples of the breadth of variation that self-generated brain activity will yield.

The above hypothesis assumes that, for the most part, the connectivity of the human brain is present at birth and "fine-tuned" during normal maturation. This view of a neurological a priori was suggested in early neurological research (Cajal 1929;

Harris 1987), with the identification by Broca of a cortical speech center and the discovery of point-to-point somatotopic maps in the motor and sensory cortices (Penfield and Rasmussen 1950) and in the thalamus (Mountcastle and Hennemann 1949, 1952).

A second organizing principle may be equally important, one that is based on the temporal rather then the spatial relationships among neurons. This temporal mapping may be viewed as a type of functional geometry (Pellioinisz and Llinás 1982). This mechanism has been difficult to study until recently since it requires the simultaneous measurement of activity from large numbers of neurons and is not a parameter usually considered in neuroscience.

3 Temporal Mapping and Cognitive Conjunction

Synchronous neuronal activation during sensory input has recently been studied in the mammalian visual cortical cells when light bars of optimal orientation and displacement rate are presented (Eckhorn et al. 1988; Crick and Koch 1990; Gray and Singer 1989). Furthermore, the components of a visual stimulus corresponding to a singular cognitive object, e.g., a line in a visual field, yield coherent gamma-band oscillations in regions of the cortex that may be as far as 7 mm(Crick and Koch 1990; Gray and Singer 1989; Singer 1993) apart or may even be in the contralateral cortex. In fact, gamma-band oscillatory activity between related cortical columns has a high correlation coefficient under such circumstances. In addition, coherent 40-Hz oscillations throughout the cortical mantle of awake human subjects has been revealed by magnetoencephalography(Llinás and Ribary 1992). These oscillations may be reset by sensory stimuli, and phase comparison revealed the presence of a 12- to 13-msec phase shift between the rostral and caudal poles of the brain (Llinás and Ribary 1992). These gamma oscillations display a high degree of spatial organization and thus may be a candidate mechanism for the production of temporal conjunction of rhythmic activity over a large ensemble of neurons.

From a neuronal point of view. the mechanism by which gamma oscillation may be generated has been studied at the level of single neurons and of neuronal circuits. For example, it has been shown that the membrane potential of sparsely spiny inhibitory neurons in cortical layer IV supports gamma frequency membrane voltage oscillation, with the mechanism for the oscillation being a sequential activation of a persistent low-threshold sodium current (Llinás and Sugimori 1980) followed by a subsequent potassium conductance (Llinás et al. 1991). The inhibitory input of these sparsely spinous interneurons onto pyramidal cells projecting to the thalamus can entrain 40-Hz oscillation in the reticular nucleus and so entrain, by rebound activation, the specific and nonspecific thalamus. This issue will be treated in the modeling part of this paper. Indeed, since the GABAergic reticular thalamic neurons project to most of the relay nuclei of the thalamus (Steriade et al. 1984), layer IV cells would indirectly make a contribution to the 40-Hz resonant oscillation in the thalamocortical network. It has recently been demonstrated that, under in vivo conditions, relay-thalamic and reticular-nucleus neurons and pyramidal cells themselves are capable of close to 40-Hz oscillation on their own, laying out in this manner the possibility for network resonance intrinsically at gamma-band frequency (Steriade et al. 1993). The ionic mechanisms underlying this oscillation are similar to those of the spiny layer IV neurons (Steriade et al. 1991).

When the interconnectivity of these nuclei is combined with the intrinsic properties of the individual neurons, a network for resonant neuronal oscillation emerges in which specific cortico-thalamo-cortical circuits would tend to resonate at gamma frequency. According to this hypothesis, neurons at the different levels, and most particularly those in the reticular nucleus, would be responsible for the synchronization of gamma oscillation in distant thalamic and cortical sites. As we will see later, these oscillations may be organized globally over the CNS, especially as it has been shown that neighboring reticular-nucleus cells may be linked by dendro-dendritic and intranuclear axon collaterals (Deschenes et al. 1985).

4 Thalamocortical Resonance and Consciousness

Based on research on the minimal temporal interval to sensory discrimination, we may conclude that consciousness is a noncontinuous event determined by synchronous activity in the thalamocortical system (Joliot et al. 1994). Since this activity is present during REM sleep (Llinás and Ribary 1993) but is not seen during non-REM sleep, we may postulate further that the resonance is modulated by the brainstem and would be given content by sensory input in the awake state and by intrinsic activity during dreaming. These studies addressed issues concerning 1) the presence of gamma-band activity during sleep and 2) the possible differences between gamma resetting in different sleep/wakefulness states.

Spontaneous magnetic activity was recorded continuously during wakefulness, delta sleep and REM sleep using a 37-channel sensor array positioned as shown in Figure 1A. Since Fourier analysis of the spontaneous, broadly filtered rhythmicity (1–200 Hz) demonstrated a large peak of activity at 40 Hz over much of the cortex, we decided that it was permissible to filter the data at gamma-band frequency (30–50 Hz). Large, coherent signals with a very high signal-to-noise ratio were typically recorded from all 37 sensors, as shown in Fig. 1B for a single, 0.6-sec epoch of global spontaneous oscillations in an awake individual.

The second set of experiments examined the responsiveness of the oscillation to an auditory stimulus during wakefulness, delta sleep and REM sleep. The stimulus comprised frequency-modulated 500-msec tone bins, triggered 100 msec after the onset of the 600-msec recording epoch; recordings were made at random intervals over about 10 minutes. In agreement with previous findings (Llinás and Ribary 1993; Galambos et al. 1981; Pantev et al. 1991), auditory stimuli produced well-defined 40-Hz oscillation during wakefulness (Fig. 1C) but no resetting was observed during delta (Fig. 1D) or REM sleep (Fig. 1E) in this or the six other subjects examined.

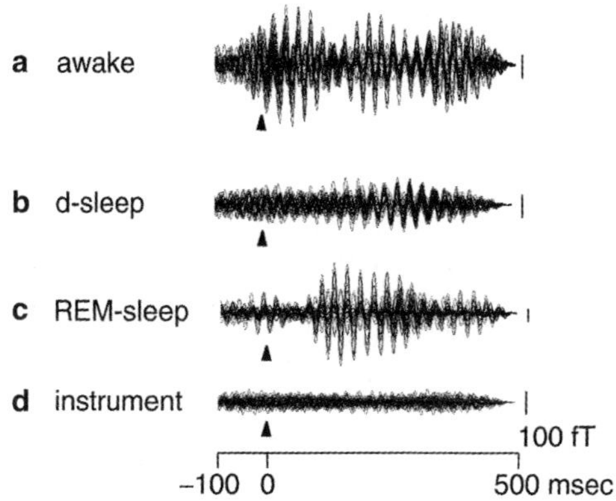

Fig. 1 The 40-Hz oscillation in wakefulness and a lack of 40-Hz reset in delta (**d**) sleep and REM sleep. In **a–d**, averaged oscillatory responses (300 epochs) following auditory stimulus (arrowheads). In A, the subject is awake and the stimulus is followed by a reset of 40-Hz activity. In **b** and **c**, the stimulus produced no resetting of the rhythm. (**d**) The noise of the system in femtotesla (fT). (Modified from Llinás and Ribary 2001)

The traces in Fig. 1 C are a superposition of the 37 traces recorded during a single 600-msec epoch. Their alignment in panel C indicates the high level of coherence of the 40-Hz activity at all the recording points following the auditory stimulus. A high level of coherence is also typical of spontaneous 40-Hz bursts.

These findings indicated that, while the awake state and the REM sleep state are electrically similar with respect to the presence of 40-Hz oscillations, a central difference remains that of the inability of sensory input to reset the 40-Hz activity during REM sleep. By contrast, during delta sleep, the amplitude of these oscillators differs from that of wakefulness and REM sleep but, as in REM sleep, there is no 40-Hz sensory response. Another significant finding is that gamma oscillations are not reset by sensory input during REM sleep, although clear evoked-potential responses indicate that the thalamo-neocortical system is accessible to sensory input (Llinás and Paré 1991; Steriade 1991). We consider this to be the central difference between dreaming and wakefulness. These data suggest that we do not perceive the external world during REM sleep because the intrinsic activity of the nervous system does not place sensory input in the context of the functional state being generated by the brain (Llinás and Paré 1991). That is, the dreaming condition is a state of hyperattentiveness to intrinsic activity in which sensory input cannot access the machinery that generates conscious experience.

An attractive possibility in considering the morphophysiological substrate is that the "nonspecific" thalamic system, particularly the intralaminar complex, plays an important part in such coincidence generation. Indeed, neurons in this complex project in a spatially continuous manner to the most superficial layers of all cortical areas, including the primary sensory cortices. This possibility is particularly attractive given that single neurons burst at 30–40-Hz Steriade et al. 1993) especially during REM sleep, which is a finding consistent with the macroscopic magnetic recordings observed in this study, and given that damage of the intralaminar system results in lethargy or coma (Facon et al. 1958; Castaigne et al. 1962).

5 Binding of Specific and Nonspecific Gamma-Band Activity as the Basis for Umwelt

A schematic of a neuronal circuit that may subserve temporal binding is presented in Figure 2. Gamma oscillations in neurons in specific thalamic nuclei (Llinás and Steriade 2006) establish cortical resonance through direct activation of pyramidal cells and feed forward inhibition through activation of 40-Hz inhibitory interneurons in layer IV (Llinás et al. 1991). These oscillations return to the thalamus via layer VI pyramidal cell axon collaterals (Steriade et al. 1990), producing thalamic feedback inhibition via the reticular nucleus (Steriade et al. 1984). A second system is illustrated on the right side of Figure 2. Here, the intralaminar nonspecific thalamic nuclei projection to cortical layers I and V and to the reticular nucleus (Penfield and Rasmussen 1950) is illustrated. Layer V pyramidal cells return oscillations to the reticular nucleus and intralaminar nuclei. The cells in this complex have been shown to oscillate at gamma-band frequency (Steriade et al. 1993) and to be capable of recursive activation.

It is also apparent from the literature that neither of these two circuits alone can generate cognition. Indeed, as stated above, damage of the nonspecific thalamus produces deep disturbances of consciousness whereas damage of specific systems produces loss of the particular modality. Although at this early stage it must be quite simple in its form, the above suggests a hypothesis regarding the overall organization of brain function. This hypothesis rests on two tenets. First, the "specific" thalamocortical system is viewed as encoding specific sensory and motor activity by the resonant thalamocortical system, which is specialized to receive such inputs (e.g., the LGN and visual cortex). The specific system is understood to comprise those nuclei, whether sensorimotor or associative, that project mainly, if not exclusively, to layer IV in the cortex. Second, following optimal activation, any such thalamocortical loop would tend to oscillate at gamma-band frequency and activity in the "specific" thalamocortical system could be easily "recognized" over the cortex by this oscillatory characteristic.

In this scheme, areas of cortical sites "peaking" at gamma-band frequency would represent the different components of the cognitive world that have reached optimal activity at that time. The problem now is the conjunction of such a fractured description into a single cognitive event. We propose that this conjunction could come about by the concurrent summation of specific and nonspecific 40-Hz activity along the radial dendritic axis of given cortical elements, that is, by coincidence detection.

In the context of this meeting, then, Umwelt is from my neuroscience perspective what the cognitive brain makes of the sensory input it receives. While the ecological universe that actually allowed and supported brain evolution is de facto an very

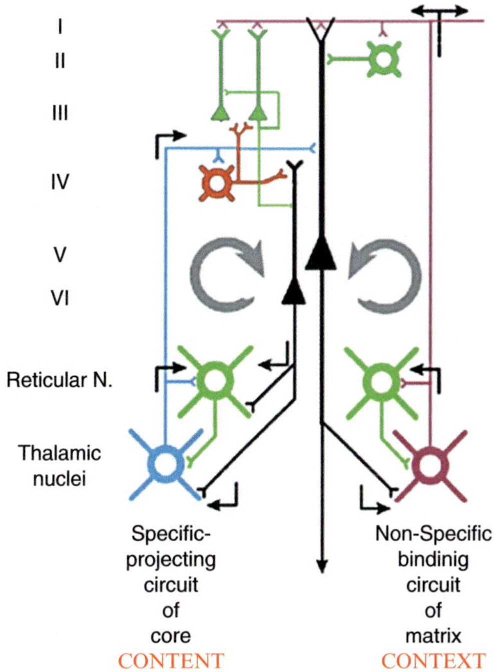

Fig. 2 Thalamocortical circuits proposed to subserve temporal binding. Diagram of two thalamocortical systems, the Content and Context networks. (Left) *CONTENT LOOP* shows specific sensory or motor thalamic nuclei neurons projecting to layer IV of the cortex, producing cortical oscillation by direct activation of cortical neurons via excitatory interneurons (red) and feed-forward inhibition via 40-Hz inhibitory interneurons (green). Collaterals of these cortical recurrent projections produce thalamic feedback inhibition via the thalamic reticular nucleus (green). The return pathway (circular arrow on the right) return this oscillation to specific- and reticularis-thalamic nuclei via layer V and VI pyramidal cells. (Right) *CONTEXT LOOP* shows nonspecific intralaminary nuclei projecting to the most superficial layer of the cortex on to pyramidal layer V neurons at their tuft, and via excitatory and inhibitory interneurons, to layer VI pyramids. Pyramidal cells return oscillation to the thalamic nuclei, establishing a resonant loop. The conjunction of the specific and nonspecific loops, driven by the content and context loops, are proposed to generate the temporal binding responsible for cognition in the sense of Umwelt

important part of this equation, it is also clear that my understanding of such an "ecological universe" can only be addressed via the cognitive experiences and deductions supported by brain function.

Acknowledgments Supported by NIH-NINCDS NS13742 to R.L.

References

Bienenstock E, Von der Malsburg C (1986) Statistical coding and short-term synaptic plasticity: a scheme for knowledge representation in the brain. In: Bienenstock, E, Fogelman R, Weisbuch G (eds) Disordered Systems and Biological Organization. Les Houches: Springer-Verlag, pp. 247–272.

Cajal SR (1929) Etude sur la neurogénese de quelques Vertébrés. Springfield, Thomas

Castaigne P, Buge A, Escourolle R, Masson M (1962) Ramollissement pédonculaire médian, tegmento-thalamique avec ophtalmoplégie et hypersomnie. Rev Neurol 106:357–367.

Crick F, Koch C (1990) Some reflections on visual awareness. Cold Spring Harbor Symp Quant Biol 55:953–962.

Deschênes M, Madariaga-Domich A, Steriade M (1985) Dendrodendritic synapses in the cat reticularis thalami nucleus: A structural basis for thalamic spindle synchronization. Brain Res 334:165–168.

Eckhorn R, Bauer R, Jordan W, Brosch M, Kruse W, Munk M, Reitbock HJ (1988) Coherent oscillations: A mechanism of feature linking in the visual cortex? Biol Cybern 60:121–130.

Edelman GM (1987) Neuronal Darwinism: the theory of neuronal group selection. New York, Basic Books.

Facon E, Steriade M, Wertheim N (1958) Hypersomnie prolongée engendrée par des lésions bilatérales due systèm activateur médial le syndrome thrombotique de la biffurcation du tronc basilaire. Rev Neurol 98:117–133.

Galambos R, Makeig S, Talmachoff PJ (1981) A 40-Hz auditory potential recorded from the human scalp. Proc Natl Acad Sci USA 78:2643–2647.

Gray CM, Singer W (1989) Stimulus-specific neuronal oscillations in orientation columns of cat visual cortex. Proc Natl Acad Sci USA 86:1698–1702.

Gray CM, Konig P, Engel AK, Singer W (1989) Oscillatory responses in cat visual cortex exhibit inter-columnar synchroniza-tion which reflects global stimulus properties. Nature 338:334–337.

Harris WA (1987) Neurogenetics. In: Adelman G (ed) Encyclopedia of neuroscience, Basel, Birkäuser, pp. 791–793.

Joliot M, Ribary U, Llinás R ((1994) Neuromagnetic oscillatory activity in the vecinity of 40 Hz coexists with cognitive temporal binding in the human brain. Proc Natl Acad Sci USA 91:11748–11751.

Kristofferson AB (1984) Quantal and deterministic timing in human duration discrimination. Ann NY Acad Sci 423:3–15.

Llinás R (1990) Intrinsic electrical properties of mammalian neurons and CNS function. Fidia Research Foundation Neuroscience Award Lectures. Vol. 4.New York, Raven Press Ltd., pp. 1–10.

Llinás R, Paré D (1991) Of dreaming and wakefulness. Neuroscience 44:521–535.

Llinás R, Pare R (1996) In: Llinas R, Churchland P The mind-brain continuum. Cambridge, MIT Press pp1–18.

Llinás RR, Ribary U (1992) Rostrocaudal scan in human brain: a global characteristic of the 40-Hz response during sensory input. In: Basar E, Bullock T (eds) Induced rhythms in the brain. Chapter 7. Boston, Birkhäuser, pp. 147–154.

Llinás R, Ribary U (1993) Coherent 40-Hz oscillation characterizes dream state in humans. Proc Natl Acad Sci USA 90:2078–2081.

Lllinás R, Ribary U (2001) Consciousness and the brain. The thalamocortical dialogue in health and disease. Ann NY Acad Sci 929:pp 166–175.

Llinás RR, Steriade M (2006) Bursting of thalamic neurons and states of vigilance. J Neurophysiol 95: 3297–3308.

Llinás R, Sugimori M (1980) Electrophysiological properties of in vitro Purkinje cell somata in mammalian cerebellar slices. J Physiol (London) 305:171–195.

Llinás RR, Grace AA, Yarom Y (1991) In vitro neurons in mammalian cortical layer 4 exhibit intrinsic activity in the 10 to 50 Hz frequency range. Proc Natl Acad Sci USA. 88:897–901.

Llinás R, Ribary U, Jeanmonod D, Kronberg E, Mitra PP (1999) Thalamo-cortical dysrhythmia: a neurological and neuropsychiatric syndrome characterized by magnetoencephalography. Proc Natl Acad Sci USA Vol. 96: 15222–15227.

Mountcastle VB, Hennemann E (1949) Pattern of tactile representation in thalamus of cat. J Neurophysiol 12:85–100.

Mountcastle VB, Hennemann E (1952) The representation of tactile sensibility in the thalamus of the monkey. J Comp Neurol 97:409–440.

Pantev C, Makeig, S, Hoke M, Galambos R, Hampson S, Gallen C (1991) Human auditory evoked gamma-band magnetic fields. Proc Natl Acad Sci USA 88:8996–9000.

Pellionisz A, Llinás RR (1982) Space-time representation in the brain. The cerebellum as a predictive space-time metric tensor. Neuroscience 7:2949–2970.

Penfield W, Rasmussen T (1950) The cerebral cortex of man. New York, MacMillan.

Singer W (1993) Synchronization of cortical activity and its putative role in information processing and learning. Ann Rev Physiol 55:349–374.

Steriade M (1991) Alertness, quiet sleep, dreaming In: Peters a. Jones EG (eds) Cerebral cortex. New York, Plenum, pp. 279–357.

Steriade M, Parent A, Hada J (1984) Thalamic projections of reticular nucleus thalami of cat: A study using retrograde transport of horseradish peroxidase and double fluorescent tracers. J Comp Neurol 229:531–547.

Steriade M, Jones EG, Llinás R (1990) Thalamic oscillations and signalling. New York, John Wiley & Sons.

Steriade M, CurróDossi R, Paré D, Oakson G (1991) Fast oscillations (20–40 Hz) in thalamocortical systems and their potentiation by mesopontine cholinergic nuclei in the cat. Proc Natl Acad Sci USA 88:4396–4400.

Steriade M, CurróDossi R, Contreras F (1993) Electrophysiological properties of intralaminar thalamocortical cells discharging rhythmic ([a]40 Hz) spike-bursts at [a]1000 Hz during waking and rapid eye movement sleep. Neuroscience 56:1–9.

Von der Malsburg C (1981) The correlation theory of brain function. Internal report, Max-Planck Institute for Biophysical Chemistry. Goettingen Germany.

Wilson JR, Friedlander MJ, Sherman SM (1984) Ultrastructural morphology of identified X- and Y-cells in the cat's lateral geniculate nucleus. Proc Royal Soc B 221:411–436.

The Brain's View of the World Depends on What it has to Know

Wolf Singer

Abstract It is argued that perception is a highly constructive process and that the way in which we perceive the world and ourselves depends on a priori knowledge. Sources of this knowledge are evolution, early developmental imprinting and life long learning processes. Much of this knowledge is implicit and therefore there is no conscious recollection of the fact that perception is determined and constrained by priors that are genetically transmitted and acquired through early experience. Moreover, these priors are adapted to the mesoscopic scale of the world in which life has evolved and therefore cognitive abilities are to be seen as the result of evolutionary and developmental adaptations to an extremely narrow segment of the world as it is known to us to date. This has far reaching consequences for epistemic considerations and perhaps also for the management of cultural conflicts. If the perception of social conditions is also dependent on priors and if these priors are acquired early during development, they will exhibit culture specific traits but will remain implicit because episodic memory develops only several years after birth (childhood amnesia). In this case subjects cannot realize that their perception of social conditions depends on idiosyncratic, culture specific priors. What is perceived will be taken as absolute truth, acquire the status of convictions and cannot be altered by arguments.

It is appropriate to begin this chapter with an epistemic caveat that is motivated by neurobiological evidence. Our perceptions and imaginations as well as our abilities to reason are constrained by the cognitive abilities of our brains, and brains, like all other organs, are the product of an evolutionary process. Hence, our brains have become adapted to the conditions of the mesoscopic world in which life has evolved. It is the world within the scale of millimeters to meters, it is the world where the laws of classical physics dominate, it is not the world described by quantum- or astrophysics. As a consequence, our cognitive functions have become adjusted to assure

W. Singer
Max Planck Institute for Brain Research, Frankfurt am Main, Germany
e-mail: singer@mpih-frankfurt.mpg.de

A. Berthoz and Y. Christen (eds.), *Neurobiology of "Umwelt": How Living Beings Perceive the World,* Research and Perspectives in Neurosciences,

survival in this mesoscopic world. Problem solving in this dangerous and poorly predictable world requires the application of pragmatic heuristics and, hence, cognitive abilities that are with all likelihood not optimized to comprehend the essence behind the perceivable phenomena or the "absolute truth" in the Kantian sense. In addition, our intuitive understanding of conditions is adapted to the mesoscopic world that is directly perceivable through our senses. This fact accounts for the tremendous difficulties we have developing intuitions for the nature of processes in the scale of elementary particles or the universe. Evolution did not endow us with sensors for processes at subatomic or cosmic scales, because they were and are completely irrelevant to our daily struggle to survive. Even more worrying is the possibility that the way in which we reason may also be constrained by adaptation to the narrow segment of the world that we can access with our specialized senses. In conclusion, it is very likely that our cognition is constrained, and to an unknown extent, and this may apply not only to primary perception but also to our way of deriving inferences from observables. If this were true, it would pose insurmountable barriers to our attempts to understand, as it would also challenge the consistency of mathematical theories and logical deductions. However, for these very reasons, we have no way to know.

1 Current Beliefs in Neuroscience

Because of the tight correlations between brain lesions and loss of function on the one hand and between neuronal activation patterns and specific cognitive or executive functions on the other, it can be inferred with high confidence that all cognitive and executive functions displayed by human beings, including all mental activities and consciousness, are based on and are the result of neural interactions rather than their cause. Furthermore, it appears that neural processes follow the known laws of nature. So far, there are no scientifically documented observations that would force us to revise this notion. It is based on the evidence that the behavior of organisms of low complexity, such as molluscs or worms, can be fully accounted for by the activity of their neurons. It is possible to establish causal relations between the spatio-temporal patterns of this activity and the respective behavior, without having to postulate any additional unknown forces, laws or modes of interaction. Furthermore, evidence from comparative studies indicates that neuronal interactions in more evolved brains continue to obey the same principles and just increase in complexity. There seems to be no ontological discontinuity in evolution. Of course, one needs to be aware that human beings are embedded in a complex socio-cultural environment and that our brains are shaped not only by our genes but also by the epigenetic influences of this socio-cultural environment - an aspect that has not been considered sufficiently in the recent past - but this concerns the specific epigenetic modifications of our brains, not the basic functional principles.

2 Cognitive Neurosciences, a Bridge to the Humanities

Since the neurosciences have begun to study the neuronal correlates of higher cognitive functions, a fascinating dialogue has been initiated within certain domains of the humanities. One of these domains is epistemology. Cognitive neuroscience explores from a third-person perspective the mechanisms that mediate our perception and the acquisition of knowledge. The long-standing discussions about the objectivity of cognition, the question of how constructive perceptual processes are or how reliable or idiosyncratic, are reconsidered on the basis of neurobiological data. Another philosophical question that neuroscientists seek to answer is related to the mind-body problem: how can mental phenomena, the immaterial entities such as the qualia of perception and social realities such as belief and value systems, emerge from the material interactions between nerve cells? These phenomena are real and profoundly affect our lives, but they have another ontological status as the neuronal processes that brought them into this world by allowing homo sapiens to initiate the evolution of cultures.

Yet another question that solicits discussions between neuroscientists and philosophers of the mind is the nature of consciousness. It is closely related to the question of the constitution of the intentional Self and the conundrum of the existence of free will. If neuronal processes are the basis and cause of all mental phenomena and if brain processes follow the laws of nature, then the principle of causality must also hold for neuronal interactions. Even though there is noise and interference, each state of the brain depends with high probability on the immediately preceding state. This is a necessary prerequisite for adapted behavior, because the brain must respond to changing conditions in a predictable way. Decisions are nothing but special brain states; hence their outcome also depends on the immediately preceding state of the brain, irrespective of whether the state variables remain unconscious or whether they enter conscious awareness and then manifest themselves as reportable feelings, thoughts and arguments. This notion has far-reaching consequences for our self-understanding and legal systems but cannot be discussed further in the present context.

3 Epistemic Considerations

One focus of this conference is the epistemic aspects of cognitive neuroscience. The majority of neurobiologists agree that perceiving is essentially reconstructing. The sparse sensory signals that we extract from the world are interpreted in a highly idiosyncratic and selective way on the basis of a huge amount of stored a priori knowledge about the world. Our brains are self-active and continuously generate knowledge and context-dependent expectancies, interpret sensory signals as a function of these inferences and present the results of these reconstructions to the workspace of consciousness. We perceive only the result of highly complex computational operations, not the operations themselves, and we experience the perceived results as if they were direct attributes of reality. What then is the neuronal basis of the a priori knowledge required for the interpretation of sensory signals? It is commonly accepted that all knowledge and the rules according to which this knowledge is processed and applied reside in the functional architecture of the brain. This contradicts the often-suggested analogy between computers and brains. Computers have processors and separate memories for program and data. In the brain, there are only neurons and connections. Both the stored knowledge and the programs for the processing of this knowledge reside in the layout of these connections and their graded efficacy. The question of the origin of stored information is thus reduced to the question: which processes determine the functional architecture of brains?

4 The Origin of Implicit Knowledge and its Role in Cognition

The most important determinant of brain architectures and hence the most important source of knowledge is, of course, evolution. What makes our brain architectures comparable is information residing in our genes. It is knowledge about the world that has been acquired by trial, error and evolutionary selection and gets expressed in the functional architecture of brains every time a new organism develops. Another important source is the epigenetic modifications of brain architecture. The human brain develops until about the age of 20. During this very protracted development, only a few new neurons are born but there is a massive proliferation of connections, and this proliferation is paralleled by a functional validation of connections that leads to the removal of newly formed pathways if they are identified as inappropriate. A large number of connections that are first formed are destroyed again to form the mature architecture of the brain. This making and breaking of connections is guided by activity, i.e., by experience. Thus, the brain adapts itself to the particular environment in which it evolves. In humans, this environment also comprises the socio-cultural context, and this is why our brains, once fully developed, differ from those of our cave-dwelling ancestors even though there can be only minor changes in the genetic dispositions. Finally, knowledge is acquired by the learning that accompanies us throughout life. This learning is based essentially on graded and lasting modifications of the coupling strength of the existing connections between neurons. It also goes along with structural changes, but these occur at the microscopic and molecular level. In the adult brain, no new long-range connections are formed and, under normal conditions, no breaking of connections occurs except at very great age.

In conclusion, evolution can be considered as a cognitive process through which the evolving organisms acquire knowledge about those features of the world whose consideration facilitates survival and reproduction. This knowledge is probably not adapted to support the understanding of the deeper structure of the world that assures coherence across scales. Knowledge acquired through evolution is stored in the genes, expressed in the functional architecture of the brain and then complemented by developmental imprinting and learning. Together, this a priori knowledge determines what and how we perceive. It defines, for example, the sensory categories according to which we assign qualities to our experiences. And these assignments are rather idiosyncratic. Thus, we arbitrarily classify electromagnetic radiation with wavelengths between 400 to 700 nanometers as light, because the photoreceptors in the eye are sensitive to this wavelength range. Radiations with slightly longer wavelengths stimulate our temperature receptors and we categorize the respective sensations as temperature. A similar arbitrariness of category boundaries is observable in other sensory domains.

A priori knowledge is also required for the definition of the nature of objects. We have clear concepts of what an object is, and the corresponding rules were worked out in great detail by the Gestalt psychologists at the beginning of the last century (see, e.g., Wertheimer 1945). Objects should be confined by continuous borders, should be solid and, if they move, all components should move in a coherent way. This definition is appropriate in the mesoscopic world but does not apply to objects at atomic or subatomic scales. If we had no a priori definition of the properties of objects, we would not be able to distinguish objects, and we would be unable to extract object-specific features from the two-dimensional brightness distribution that complex scenes generate on the retina. The principles according to which we associate features with objects reside in the circuitry of the visual system. Similarly, the learning rules according to which we establish associations are implemented by specific molecular mechanisms that have been preserved nearly unchanged since the evolution of primitive nervous systems. A particularly interesting epistemic problem would arise if the same idiosyncratic adaptation to the special conditions of the mesoscopic world also held for the way we reason, infer and assign values to certain activity patterns and not others. The respective rules must also reside in the functional architecture of the brain and, as mentioned, this architecture is the result of adaptation to the mesoscopic scale and hence to a special segment of the world. In this case, the way in which we reason may not be generalizable across scales.

In this context, it needs to be emphasized that evolution is a very conservative process. Once an invention has been made that works, it tends to be conserved unless there is a major change in conditions that makes this invention obsolete or maladapted. This is why our nerve cells function in exactly the same way as those of snails, and why the same learning rules are implemented and the same mechanisms of signal transduction apply. Also the development of structures followed a very conservative path. Since the first appearance of the cerebral cortex, the six-layered sheet of nerve cells that covers the hemispheres of the brain, no new structures have emerged. There is just more of the same and this increase in hardware makes all the difference. It is the addition of a few more cortical areas that seems to make the difference between the brain of a human being and that of our nearest neighbors, the great apes. Apparently it is only this addition of processing substrate and the associated gain of complexity that are responsible for the difference between animals and humans, between species that failed and those that succeeded to promote cultural evolution with all its far-reaching consequences. In this context, however, one needs to consider that cultural evolution created a socio-cultural environment of ever-increasing complexity that in turn contributed to the epigenetic shaping of

Fig. 1 Mold used to produce candies that illustrate how a priori knowledge structures our perceptions. On the left side one sees the front aspect of the mold, with the concavities, and on the right, the rear side with the corresponding convex protrusions. In reality, both pictures show the front aspect, but one picture is rotated by 180°

brain architectures. Thus, even if the genetically determined layout of brain architectures has changed only little since the beginning of human civilizations, the features that can be modified by epigenetic shaping are likely to have undergone major modifications.

Here are two examples that illustrate how a priori knowledge structures our perceptions (Fig. 1). The object in Figure 1 is a mold used to produce candies. On the left side one sees the front aspect of the mold, with the concavities, and on the right, the rear side with the corresponding convex protrusions. In reality, both pictures show the front aspect, but one picture is rotated by 180°. The reason for these very different perceptions is that the brain makes the a priori assumption that light comes from above. In this case, contours that have shadows above need to be interpreted as concave and those with shadows below as convex. Thus, an implicit assumption determines what we perceive. Somehow this assumption is implemented in the processing architecture of the visual cortex, but we are not aware of it. Another really striking example is shown in Figure 2. It is hard to believe, but surfaces A and B have exactly the same luminance. They appear so different because the brain sees the shadow that is caused by the cylinder on the right. Even though the amounts of light reflected from surfaces A and B and impinging on the retina are exactly the same, the brain interprets the brightness of the two surfaces as different because it infers the following: given that there is a shadow, surface B must be brighter

Fig. 2 It is hard to believe, but surfaces A and B have exactly the same luminance. They appear so different because the brain sees the shadow that is caused by the cylinder on the right

than surface A, which has no shadow on it, in order to reflect the same amount of light. Thus, the brain "computes" the inferred brightness of the surfaces but we are not aware of these computations. We just perceive the result and take it as real, i.e., we see B is much brighter than A. One could spend hours with examples that indicate that the brain generates inferences that we are not aware of, that it is permanently reconstructing the world according to a priori knowledge and that we, as perceiving subjects, have to take for granted what the system finally offers us as conscious experience. As expected, this is not only the case with specially designed psycho-physical experiments but it is also an essential feature of all our perceptual processes. As far as the perception of the material world is concerned, this poses no major problems. It is even a well-adapted strategy because it allows the organism to preserve perceptual constancy by evaluating relations rather than absolute values. Humans evolved in similar environments and hence share the same cognitive mechanisms. We perceive the world in the same way and tend to agree on most of its attributes.

However, problems would arise if the perception of socio-cultural attributes also depended on acquired, implicit schemata. Human cultures exhibit a high degree of diversity. If the perception of social realities is also determined by implicit knowledge, acquired through early imprinting and education, it is to be expected that members of different cultures would perceive attributes of the socio-cultural world in radically different ways without being aware of this fact. They would be bound to take for granted what they experience and, even if they knew that any experience was biased by cognitive schemata and complex inferences, they would not be able to perceive the world otherwise – just as it does not help to know about shadows and reflectance to make the squares in the checkerboard appear equally bright. This dependence of perception on imprintable, implicit knowledge is perhaps at the root of the many dramatic conflicts between different ethnical groupings. If the primary perception of particular socio-cultural traits differs because of differently primed cognitive schemata, discussions on wrong or right are bound to be fruitless. Unlike in the material world, there is no way – at least not yet – to obtain independent evidence for the identification of perceptual illusions in the social domain.

5 Where is the Observer in the Brain?

As mentioned above, neurobiologists assume that all mental functions including phenomenal awareness and self-consciousness are based on neuronal processes. Here we have a severe explanatory gap. We encounter extreme difficulties when we attempt to explain how exactly the qualia of our subjective experience emerge from neuronal interactions. We describe the two phenomena in different language systems and are lacking satisfactory bridging theories. We can show with experiments that there is only a minor difference between imagining something and perceiving it in reality. If one investigates subjects in the scanner and asks them to imagine, for example, a rotating wheel, one observes a characteristic distribution of activities in pre-frontal brain regions that need to be engaged to activate the read-out of stored information. In addition, there is activity in the visual areas that most likely corresponds to the read-out of the memorized images of rotating wheels. If one then asks the subject to open his eyes and to look at the image of a rotating wheel, only a little additional activation occurs and it is confined to the primary visual area of the cerebral cortex, where the activity from the eyes is first processed (Goebel et al. 1998). Thus, the difference between the activity patterns associated with imagination and real perception is minimal. This is also why the brain's threshold for hallucinations is so low that sleep deprivation or fever can blur the distinction between imagery and perception. The best-examined hallucinatory states are those occurring with schizophrenia. We investigated schizophrenic patients who suffered from verbal hallucinations and we registered the activity in the auditory cortex of the speech-competent hemisphere (Dierks et al. 1999). Every time patients reported by a button press in the scanner that they heard voices that they attributed to a real speaker, there was activity in the primary auditory cortex. This activity had about the same amplitude as the responses to voices presented through earphones. When healthy subjects imagine silent speech, they activate their speech centers but not the primary auditory cortex. In the patients, this self-generated activity in speech centers apparently propagated all the way down to the primary auditory cortex, making it impossible for the subjects to distinguish between self-generated and evoked activity.

How then is it possible that we can normally distinguish between our imaginations and our perceptions of the outer world, and that we do not only perceive but can in addition be aware of the fact that we perceive? How can we explain, in neuronal terms phenomenal awareness, the ability to be aware of one's cognitive operations? There are as yet no satisfactory answers to these questions. However, some hints can be found by considering the evolutionary development of brains. In less-evolved brains, the paths from sensory to executive areas of the cerebral cortex are short. As evolution proceeds and brains become more and more complex, one observes the addition of new cortical areas that are no longer connected to the periphery, neither on the executive nor on the receptive side. They receive their information, their input, exclusively from the output of the phylogenetically older areas (see, e.g., Krubitzer 1995). This process is iterative, with more and more areas added that communicate only with other areas. A neuron located in these more recently added areas is connected exclusively with other partners in the cerebral cortex. Evidence indicates that all cortical areas, the old and the more recent ones, operate according to the same principles, because they have the same intrinsic organization. Thus, the more recent areas process the output of the older areas according to the same principles as these process the signals from the sense organs. This iteration should allow the creation of meta-representations, representations of representations of representations, which is perhaps the structural basis that allows us to subject to cognitive operations not only signals from the outer world but also those resulting from internal processes, to realize various levels of meta-cognition and to thereby become aware of our own perceptual functions.

6 The Distributed Organization of the Brain and Wrong Intuitions

A conundrum closely related to the question of the nature of the "observer" and a challenge to both brain research and our intuitions is the distributed organization of our brains. The neurobiological evidence accumulated over the last decades has led to radical changes in our views of the brain. In early days intuition and introspection were the major sources of knowledge for the formulation of hypotheses about the organization of the brain. Now we learn that these intuitions are in drastic conflict with the evidence provided by scientific investigations, raising the interesting question of why the brain is so agnostic to its own organization. We do not feel our brain, have little intuitive insight into its processes and are surprised to find that it works according to principles that differ substantially from what we thought. Intuition suggests to us that somewhere in the brain there ought to be a convergence center where all information is coming together to be amenable to coherent interpretations of the world. This would be the site where perception takes place, where the intentional agent is active, where decisions are reached, where plans are developed and where the Self is seated. We assume hierarchical structures and we also recreate them in the social and economical world, though probably not always to our advantage because they may be maladapted when systems become very complex.

The reality of our brains is very different. The cerebral cortex comprises a large number of different areas that, depending on their input, accomplish different functions but use similar computational algorithms (see, e.g., Sporns et al. 2004). Thus, the format of exchangeable information is always the same and communication among cortical areas can capitalize on this lingua franca, which is a necessary prerequisite for generalization, abstraction, symbolic encoding and, last but not least, for the constitution of the unity of consciousness. The surprising finding is that the connections linking these areas provide only little evidence of serial processing in strictly hierarchical architectures. Rather, the connectivity scheme is dominated by principles of parallelity, reciprocity and distributedness. Thus, neurons located in the visual cortex can talk directly to neurons in the limbic system or in executive areas and most of these interactions are reciprocal. This meshwork of connections is extraordinarily dense and complex but far from random. It is highly structured and has properties of so called "small world networks" (Yu et al. 2008). This architecture is the hardware realization of the programs according to which brains process information and it is also the basis of stored knowledge.

7 Distributed Representations

Imagine one perceives a barking dog, touches its fur and judges the dog as friendly. In this case, all visual areas will be active and participate in the identification of the dog; the same holds for the tactile areas, which analyze the texture of its fur, the auditory areas, which decode the barking, and the limbic areas, which add the emotional connotations. There is no single locus for the representation of the integrated percept of this dog. Rather, the representation consists of a complex spatio-temporal pattern of distributed neural activity. The brain lacks the postulated singular convergence center, which Descartes actually searched for in cow brains and thought he found in the pineal organ, because it occurs only once like the hypophysis. In reality, however, there is no such center, no observer, no coordinator; there is no identifiable seat of the conscious, intentional self. The brain is a distributed system that self-organizes and produces all those astonishing phenomena that we as observers attribute to the person, the Self. The question then arises: why is our intuition so wrong? My suspicion is that the brain, even though it exhibits non-linear dynamics, is tuned to assume that the processes to be analyzed are linear in order to be able to make reasonable predictions. However, if the brain assumes the same concerning its own functioning, if it assumes that it executes mainly linear operations, it is bound to postulate a mover. The reason is that linear systems cannot by themselves produce all the remarkable functions that we observe; they cannot be creative, open towards the future and intentional. But this interpretation is, of course, a speculation.

8 The Binding Problem

Self-organizing, distributed and goal-directed systems need efficient and flexible mechanisms to coordinate and bind in a context-dependent way the many distributed local processes into coherent wholes. One way to bind distributed results is convergence in devoted anatomical circuits. If the messages encoded by units A and B are to be bound, it suffices to connect their outputs with a third unit, C, and then to select appropriate thresholds for unit C, so that C is only active when A and B are active at the same time. In this way, relations can be evaluated in rigid, anatomical architectures and expressed by the responses of conjunction-specific neurons. The brain uses this strategy, but because of its rigidity and inflexibility, this strategy can be used only for the encoding of frequently occurring stereotyped relations (for a detailed discussion see Singer 1999; von der Malsburg 1999; Gray 1999). The alternative is to express relations by dynamic signatures, so that the representation remains distributed but functions as a coherent whole, a strategy called assembly coding (Hebb 1949). This coding strategy is much more flexible because it shares features with language systems. With 26 letters one can write the world literature, simply by re-combining the letters in a flexible way. With 10^{11} neurons, each having the role of a symbol, and a flexible recombination mechanism, a virtually infinite number of different distributed representations can be formed. The representations of novel objects, of the ever-changing constellations of real world conditions and of adaptive motor responses are thus implemented best by dynamically configurated assemblies. However, in assembly coding, one needs a code that defines from instance to instance which subset of the myriads of the active neurons actually contributes to a particular representation. As there will always be several coexisting assemblies, an unambiguous signal is needed that tells the rest of the brain which neurons are actually bound together in an assembly. In essence, neurons supporting assembly codes have to convey two messages in parallel: first they have to signal whether the feature or combination of elementary features for which they serve as symbol is present and, second they have to indicate, in parallel, with which other neurons they are actually collaborating to form the coherent whole to which they contribute their encoded information. There is common agreement that they signal the presence of their symbol, of their feature, by increasing the frequency of their discharges. The more they become active, the more reliable this message is. Following a discovery made in our lab in Frankfurt in 1986 (Gray and Singer 1989; Gray et al. 1989), we pursued the hypothesis that the signature for the *relatedness* of the cells belonging to an assembly is the precise synchronization of the individual discharges. The required precision is in the range of milliseconds to allow the definition of relations with the necessary temporal resolution, which is required to reconfigure assemblies at a rapid pace. The technical details that suggest that synchronization is an excellent tag for the definition of relations cannot be discussed here but have been summarized in numerous reviews (Singer and Gray 1995; Singer 1999; Fries et al. 2007). Intuitively, it appears as obvious that events that happen simultaneously are easily bound together, and there are indeed mechanisms that render neurons particularly susceptible to synchronous, i.e., coincident inputs and to promote permanent association of temporally contingent events by learning (Markram et al. 1997; Stuart and Häusser 2001; Wespatat et al. 2004).

9 The Role of Oscillatory Synchronized Activity

Since the discovery of stimulus-dependent synchronization of oscillatory responses in the visual cortex (Gray and Singer 1989), many laboratories have joined the search for its functional implications. A major prerequisite for those studies is to sample simultaneously the responses of at least two neurons, preferably of as many as possible, because otherwise temporal relations cannot be assessed. In this context, it is noteworthy that until recently we recorded from only one neuron at a time and related the firing of these isolated cells to stimuli or behavior to identify their functional properties. This method precluded analysis of relations and hence the identification of functionally bound assemblies. If one considers the complexity of the system, it is obvious that even multisite recordings have their limits. We are, despite all progress, still at the very beginning of understanding the brain processes underlying higher cognitive functions but we seem to know how to proceed. We have decades ahead of us in which we shall need more and more theoretical approaches and new technology, but we know where to go. Since its discovery, synchronization of oscillatory activity has become a candidate mechanism for many different functions. Dynamic binding, the flexible definition of relations, is only one of them. Synchronization seems also to be involved in attentional mechanisms that select activity for further processing by raising the saliency of responses through synchronization (Fries et al. 2001a). Synchronization also appears to serve the read-out of information that is stored in the connectivity (Fries et al. 2001b), and it may be used to bind different sub-systems together, such as sensory and motor systems (Roelfsema et al. 1997; Schoffelen et al. 2005). Evidence also indicates that it serves the selective routing of signals across the highly interconnected networks of the cerebral cortex (Fries et al. 2007). The mechanism resembles the tuning of a radio to the frequency of a certain station, thereby allowing brain centers to send a message with high selectivity from point A to point B. This selective routing is a very difficult problem in a highly connected system and is apparently solved by synchronization. There are also indications that entrainment into coherent oscillations plays a role in the storage and maintenance of information in short-term memory (Tallon-Baudry et al. 1998, 1999, 2004), and finally large scale synchronization appears to be a prerequisite for signals to have access to conscious processing (Varela et al. 2001; Melloni et al. 2007).

These more recent insights into the nature of cognitive processes reveal the limitations of our intuition. It is difficult for us to imagine the structure of distributed representations because the essential information is encoded in non-stationary, spatio-temporal relations among large numbers of neurons rather than in the responses of individual neurons. In principle, one can measure these patterns by registering the activity of large numbers of neurons simultaneously, but then one has to use advanced mathematical methods to describe those patterns. These mathematical methods result in abstract descriptions of vectors in a high-dimensional state space. They are not tangible and difficult to represent in two or three dimensions; they are essentially lists of numbers or systems of equations. For our intuition, such explanations bear little resemblance with a percept, a thought or a feeling. This epistemic problem will in all likelihood become even more accentuated in the future because descriptions of neural states will become more and more abstract, more and more mathematical, more and more intuitively implausible, not the least because the brain appears to be a non-linear system but its cognitive mechanisms have not been prepared by evolution to deal with non-linear dynamics.

References

Dierks T, Linden DEJ, Jandl M, Formisano E, Goebel R, Lanfermann H, Singer W (1999) Activation of Heschl's gyrus during auditory hallucinations. Neuron 22: 615–621

Fries P, Reynolds JH, Rorie AE, Desimone R (2001a) Modulation of oscillatory neuronal synchronization by selective visual attention. Science 291: 1560–1563

Fries P, Neuenschwander S, Engel AK, Goebel R, Singer W (2001b) Rapid feature selective neuronal synchronization through correlated latency shifting. Nature Neurosci 4: 194–200

Fries P, Nikolic D, Singer W (2007) The gamma cycle. Trends Neurosci 30: 309–316

Goebel R, Khorram-Sefat D, Muckli L, Hacker H, Singer W (1998) The constructive nature of vision: direct evidence from functional magnetic resonance imaging studies of apparent motion and motion imagery. Eur J Neurosci 10: 1563–1573

Gray CM (1999) The temporal correlation hypothesis of visual feature integration: Still alive and well. Neuron 24: 31–47

Gray CM, Singer W (1989) Stimulus-specific neuronal oscillations in orientation columns of cat visual cortex. Proc Natl Acad Sci USA 86: 1698–1702

Gray CM, König P, Engel AK, Singer W (1989) Oscillatory responses in cat visual cortex exhibit inter-columnar synchronization which reflects global stimulus properties. Nature 338: 334–337

Hebb DO (1949) The organization of behavior. John Wiley & Sons, New York

Krubitzer L (1995) The organization of neocortex in mammals: are species differences really so different? Trends Neurosci 18: 408–417

Markram H, Lübke J, Frotscher M, Sakmann B (1997) Regulation of synaptic efficacy by coincidence of postsynaptic APs and EPSPs. Science 275: 213–215

Melloni L, Molina C, Pena M, Torres D, Singer W, Rodriguez E (2007) Synchronization of neural activity across cortical areas correlates with conscious perception. J Neurosci 27: 2858–2865

Roelfsema PR, Engel AK, König P, Singer W (1997) Visuomotor integration is associated with zero time-lag synchronization among cortical areas. Nature 385: 157–161

Schoffelen J-M, Oostenveld R, Fries P (2005) Neuronal coherence as a mechanism of effective corticospinal interaction. Science 308: 111–113

Singer W (1999) Neuronal synchrony: A versatile code for the definition of relations? Neuron 24: 49–65

Singer W, Gray CM (1995) Visual feature integration and the temporal correlation hypothesis. Annu Rev Neurosci 18: 555–586

Sporns O, Chialvo DR, Kaiser M, Hilgetag CC (2004) Organization, development and function of complex brain networks. Trends Cogn Sci 8: 418–425

Stuart GJ, Häusser M (2001) Dendritic coincidence detection of EPSPs and action potentials. Nature Neurosci 4: 63–71

Tallon-Baudry C, Bertrand O, Peronnet F, Pernier J (1998) Induced γ-band activity during the delay of a visual short-term memory task in humans. J Neurosci 18: 4244–4254

Tallon-Baudry C, Kreiter AK, Bertrand O (1999) Sustained and transient oscillatory responses in the gamma and beta bands in a visual short-term memory task in humans. Visual Neurosci 16: 449–459

Tallon-Baudry C, Mandon S, Freiwald WA, Kreiter AK (2004) Oscillatory synchrony in the monkey temporal lobe correlates with performance in a visual short-term memory task. Cereb Cortex 14: 713–720

Varela F, Lachaux J-P, Rodriguez E, Martinerie J (2001) The brainweb: phase synchronization and large-scale integration. Nature Rev Neurosci 2: 229–239

von der Malsburg C (1999) The what and why of binding: The modeler's perspective. Neuron 24: 95–104

Wertheimer M (1945) Productive thinking. Harper, New York

Wespatat V, Tennigkeit F, Singer W (2004) Phase sensitivity of synaptic modifications in oscillating cells of rat visual cortex. J Neurosci 24: 9067–9075

Yu S, Huang D, Singer W, Nikolic D (2008) A small world of neural synchrony. Cereb Cortex. Advance access doi: 10.1093/cercor/bhn133

The Biology of Variations in Mammalian Color Vision

Gerald H. Jacobs

> *"... are you quite certain that the several colors appear to a dog or to any animal whatever as they appear to you?"*
>
> Plato's Theaetetus (369-367 BCE)

Abstract Although the idea clearly has a long history, the nineteenth-century English naturalist John Lubbock was among the first to provide a compelling experimental demonstration that the visual worlds of other animals can be distinctly different from those experienced by humans. Over the years since that time, an immense body of literature has been accumulated that details how these differences arise. We now know that the vast majority of all vertebrates possess some capacity for color vision—the ability to disambiguate variations in the wavelength and in the intensity of light across the visual scene. The past three decades have witnessed striking progress in our understanding of how color vision works. Key to this progress has been the detailing of the linkages between opsin genes, photopigments, and color vision. This chapter summarizes this work, focusing particular on the issue of primate color vision. In addition to helping reveal the machinery and evolution of color vision variation, this line of research also provides insights into the real-world utility of color vision.

1 Introduction

While philosophers continue to engage in protracted debates about the locus of color, whether it inheres in objects or is produced subjectively (cf., Hilbert 1987; Hardin 1988; Byrne and Hilbert 2003), most vision scientists hold the view that

---

G.H. Jacobs
Neuroscience Research Institute, University of California, Santa Barbara, CA 93106 USA
e-mail: jacobs@psych.ucsb.edu

A. Berthoz and Y. Christen (eds.), *Neurobiology of "Umwelt": How Living Beings Perceive the World,* Research and Perspectives in Neurosciences,

color is a perceptual construction based on an analysis of the spectral reflectance of objects in the photic environment by an active eye and nervous system. Isaac Newton explicitly framed this idea with his celebrated observation, "for the Rays to speak properly are not coloured, in them there is nothing else but a certain Power or Disposition to Stir up a Sensation of this or that Colour" (Newton 1730). Such an active construction of color necessarily allows for individual variation, and clear evidence for such variation was first conclusively established through a series of reports describing how color vision varies among people. Most notable among these were the rigorous self observations made in 1794 by the chemist John Dalton, which led him to conclude, "I am at present engaged in a very curious investigation: -I discovered last summer with certainty, that colours appear very different to me to what they do to others…" (reprinted in Dalton 1924). Dalton went on to show that his color perceptions were not unique, but rather they were shared in common with a small number of other males. The modern diagnosis is that Dalton had a common form of congenital defective color vision called deuteranopia (Hunt et al. 1995).

Given that the construction of color rests on organizational details of nervous systems, it follows that color for other species is apt to be quite different from what it is for humans. That idea was substantiated in early observations made by John Lubbock, an English naturalist. Lubbock provided a direct experimental demonstration that ants are sensitive to ultraviolet lights, radiations invisible to humans, and that result and others led him to the following thought, "It is, I think, generally assumed not only that the world really exists as we see it, but that it appears to animals pretty much as it does to us. A little consideration, however, is sufficient to show that this is very far from certain, or even probable" (Lubbock 1882).

Although it is not clear that von Uexküll was aware of all these earlier observations on color, his concept of "Umwelt," as describing the worlds that animals uniquely inhabit, is certainly in line with a subjectivist view (von Uexküll 1934). To the modern reader, von Uexküll's idea that you can capture these worlds as seen through human eyes seems often quite fanciful; for example, at one point he sequentially renders a photograph of a small village as it would purportedly be seen through the eyes of flies and of molluscs. That weakness in his work was evident even to his contemporaries. In the words of the psychologist, K. S. Lashley, "although the descriptions of the animals' world give the reader a feeling for their experience, this empathy is illusory and sometimes misleading" (Lashley 1934).

In recent years great progress has been made in revealing the early-stage mechanisms in the visual system through which color begins to be realized. Insights derived from this work yield particularly strong evidence as to how variations in animal color vision arise. This work is also beginning to provide access to the potentially more challenging question of why such variations may exist. The comments that follow focus on color vision and its mechanisms in mammals, with particular attention to the issue of primate color vision.

2 Opsin Genes and Photopigments

The influence of photic energy on the visual system is mediated by photosensitive pigments sequestered in receptor cells that are arrayed across the retina in a manner that permits a dense spatial sampling of the retinal image. Absorption of light by these pigments initiates a rapid cascade of changes, first in photoreceptors (cones in the case of vertebrate color vision) and then subsequently in the neurons of the primary visual system in a complex process that eventuates in visual perception. Unabsorbed light cannot contribute to vision, so photopigments play a crucial role in both limiting and defining the visual world.

The efficiency of photon absorption by photopigments varies as a function of wavelength (Fig. 1a). These absorption spectra have a common shape when appropriately scaled and thus pigments are conveniently characterized according to the wavelength at which they absorb maximally (λ_{max}). Variations in the spectral positioning of photopigments depend on amino acid dimorphisms at a small number of sites in the photopigment protein, opsin (Fig. 1b). In turn, these proteins are specified by single genes so that, ultimately, the spectral tuning of photopigments reflects variations at a small number of critical sites in the primary structure of the opsin genes (Neitz et al. 1991; Merbs and Nathans 1993; Asenjo et al. 1994).

The presence of color vision is technically defined as the ability of an animal to discriminate differences in the wavelengths of lights irrespective of any variations in their relative intensities (Jacobs 1993; Kelber et al. 2003). Note that the definition is based solely on securing evidence of successful discrimination behavior without reference to appearance or other qualitative features associated with the perception of color. Given that restriction, it follows that, at minimum, two organizational features

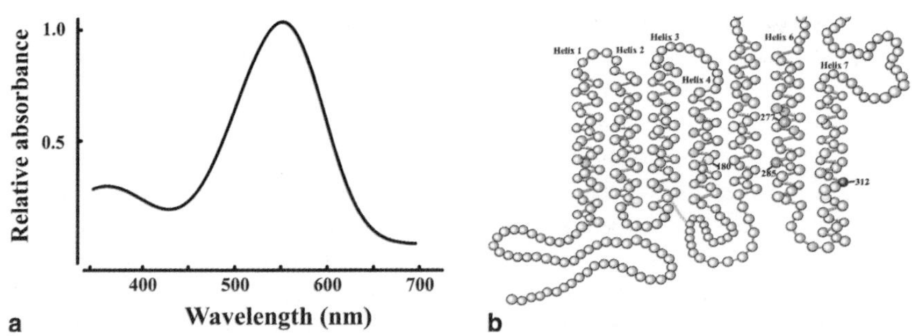

Fig. 1 **a** Spectral absorption function for a typical mammalian cone photopigment. In plots like this, the height of the curve indicates the probability that light will be absorbed as a function of wavelength. The pigment depicted here has peak absorbance (λ_{max}) at 560 nm. **b** Schematic representation of the M/L primate cone opsin molecule. The 364 amino acids in this molecule are depicted as the small circles. As suggested by this model, the opsin molecule threads its way through the photoreceptor membrane in a series of seven linked helical arrays. The three amino acid residues that are principally responsible for shifting the spectral absorption of this pigment are numbered 180, 277, and 285. The chromophore (in this case, retinal 1) is covalently bound to the opsin at residue 312

of visual systems must be met to allow an animal to have color vision (Solomon and Lennie 2007). First, there need to be two or more types of cone pigment having different absorption spectra present in the retina and, second, the nervous system must be so organized that the absorption of light by these different classes of pigment can be contrasted.

3 Evolution of Photopigments

Since the human opsin genes were first isolated and sequenced some 20 years ago (Nathans et al. 1986), hundreds of additional opsin genes from many different species have been studied. Among other useful outcomes, this effort allows broad-scale comparisons of gene structure that in turn provides insights into the evolution of these critical proteins. All vertebrate visual pigment genes are believed to fall into five paralogous (i.e., sharing a common evolutionary ancestry) families (Fig. 2). One group (Rh1) codes for all of the photopigments found in rods; the others yield pigments expressed in cones (Hisatomi and Tokunaga 2002; Jacobs and Rowe 2004). The four families of cone opsin genes (LWS, Rh2, SWS2, and SWS1) specify a large number of pigment types that, across all vertebrates, span the spectrum, with λ_{max}

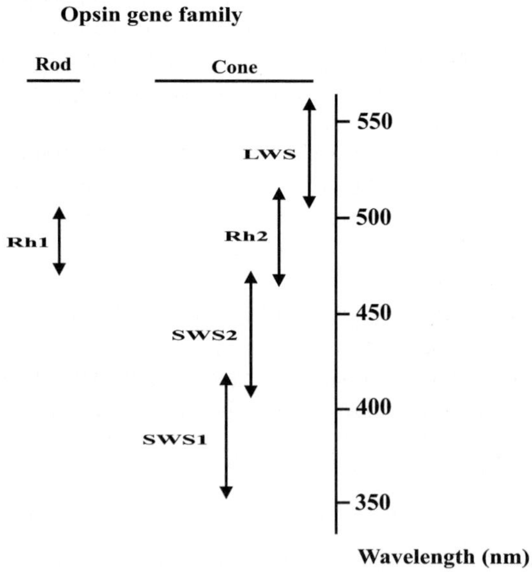

Fig. 2 Five opsin gene families specify all the photoreceptor opsins found in vertebrate retinas. One of these (Rh1) is associated with photopigments found in rods; the other four specify cone pigments. The extent of the vertical line indicates the range of λ_{max} values for retinal 1-based pigments specified by each of the gene families. As described in the text, members of the Rh2 and SWS2 opsin gene families are not found in eutherian mammals

ranging from ~360 nm to 620 nm. An important discovery was that, although cone-pigment products from all four of the gene families are represented in numerous species of birds, fishes, reptiles and amphibians, that is not the case for mammals. With the exception of a few Australian marsupials and monotremes (Arrese et al. 2002; Davies et al. 2007), mammals exclusively derive their cone pigments from only two of these gene families (Yokoyama and Radlwimmer 1998). In mammals, genes drawn from the LWS family are located on the X-chromosome whereas the SWS1 genes are autosomal, all being mapped to chromosome 5. This essential difference between mammals and other vertebrates is interpreted as reflecting the fact that representation of two of the cone opsin gene families was lost sometime during the early history of mammals, very likely when mammals are believed to have occupied nocturnal niches and when vision was consequently less important than other sensory capacities.

There are strong predictive linkages between the number of types of cone pigment and the dimensionality of color vision. Animals that have only a single type of cone pigment lack color vision (technically, they are called monochromatic). The presence of two types of cone pigment supports a single dimension of color vision - dichromatic color vision - a condition quantitatively similar to that experienced by John Dalton. Three classes of cone pigment allow two dimensions of color vision, a condition termed trichromatic, and so on to potentially higher dimensions of color vision (tetrachromatic, etc). A consequence of this linkage between the number of cone types and behavior is that it makes it possible to infer some features of color vision from knowledge of opsin genes (and vice versa) and thus derive hypotheses about color vision in animals that are not available for direct study. Importantly, this can be accomplished for ancestral as well as contemporary species. Such efforts suggest that cone opsins appeared very early in the evolution of animal life and that, subsequently, perhaps as much as 500 to 800 mya, two separate families of opsin genes emerged (Neitz et al. 2001). That step would have permitted animals to have two types of cone pigments and from this they could potentially acquire a rudimentary color vision capacity. Further comparative gene sequence analysis suggests that all four of the cone opsin gene families had emerged by the early Cambrian period (~540 mya) and, consequently, one or another form of color vision seems likely to have been characteristic of the visual worlds of vertebrates throughout their history (Collin and Trezise 2004). As noted, most contemporary mammals have two types of cone pigment supporting some form of dichromatic color vision (Jacobs 1993). Primates offer a series of intriguing exceptions to this rule that have been revealed during the past 25 years.

4 Evolution of Primate Cone Pigments and Color Vision

It has been known since the seventeenth century that most humans have trichromatic color vision, but it wasn't until the middle of the last century that it was convincingly established that this capacity derives from the presence of three classes of cone pigment (Brown and Wald 1963). These three photopigments are arrayed across the visible spectrum and are conventionally referred to as short- (S), middle- (M) and long- (L) wavelength sensitive (Fig. 3a). As in other mammals, the human S pigment is derived from the SWS1 opsin gene family. The opsin proteins of both the M and L pigments are specified by members of the LWS opsin gene family. The latter are positioned in tandem array on the q-arm of the X-chromosome. The M and L opsin genes share nearly identical sequences (~96%), implying that these two genes diverged recently, whereas similar sequence comparisons show that these two are in turn much more distantly related to the gene specifying the S-cone opsin (Nathans et al. 1986). The essential difference between human color vision and that of other non-primate mammals was first interpreted as resulting from an X-chromosome opsin gene duplication and subsequent diversification that must have occurred near the base of the catarrhine (a group that includes humans, apes and

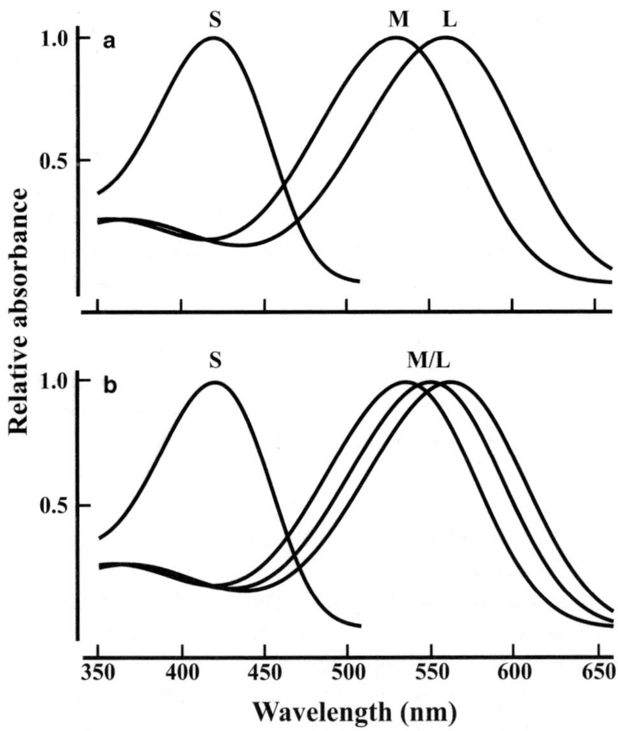

Fig. 3 Spectral absorption curves for primate cone photopigments. Panel **a** illustrates the arrangement found in humans and in other catarrhine primates. The three pigments are typically designated as S, M, and L; the former is specified by a gene from the SWS1 family, the latter two by genes from the LWS opsin gene family. Panel **b** depicts the cone pigments found in typical platyrrhine monkeys. In addition to an S pigment that is common to all individuals, these species have three polymorphic versions of the M/L pigment. As explained in the text, the retinas of individual monkeys may contain any of these three pigments (supporting dichromatic color vision) or any pair of the three, leading to trichromatic color vision

Old World monkeys) radiation, some 30 to 40 mya (Nathans et al. 1986). In strong support of that conclusion, all catarrhine primates are believed to share their three classes of cone pigment (Bowmaker et al. 1991; Jacobs and Deegan II 1999). In large measure all of these primates also seem to have very similar trichromatic color vision capacities. The conventional interpretation of these facts is that catarrhine primates were able to escape the dichromatic color vision confines within which the vision of most mammals is constrained by means of a timely duplication of the X-chromosome opsin gene.

Studies of platyrrhine (New World) monkeys reveal a quite different picture, one that has proven particularly helpful in advancing our understanding of color vision and its evolution. Unlike the catarrhines, most platyrrhine monkeys have X-chromosome opsin gene polymorphisms (for a recent review, see Jacobs 2007). In most cases there are three allelic versions of the M/L opsin gene, each of which codes for a pigment with a unique absorption spectrum (Fig. 3b). The resulting cone pigment array then varies among individual monkeys such that female monkeys that are homozygous at the X-chromosome opsin gene locus and all males have any one of the three versions of the M/L opsin gene and, in conjunction with the S-cone opsin gene common to all individuals, produce two types of cone pigment supporting one or another form of dichromatic color vision. Heterozygous females, on the other hand, have different versions of the two genes on their two X-chromosomes and, through early random X-chromosome inactivation, these female monkeys develop cones containing different M/L pigments. Direct tests of color vision reveal that the heterozygous females indeed have trichromatic color vision (Jacobs 1984; Jacobs et al. 1987; Tovee et al. 1992). The result is that these platyrrhine monkey species consist of subgroups of animals with each having one or the other of six distinct color-vision phenotypes: three forms of trichromacy and three forms of dichromacy.

There are two intriguing exceptions to this pattern of polymorphic color vision in platyrrhine monkeys. One genus (*Aotus*) lacks the L/M opsin gene polymorphism common to most platyrrhines (Jacobs et al. 1993) and, in addition, the S-cone opsin gene found in this monkey has fatal mutational changes that obviate expression of S-cone photopigment (Jacobs et al. 1996a). The retina of the *Aotus* monkey, the only nocturnal anthropoid, thus contains only a single type of cone pigment and, perforce, they have no color vision. The other exceptions to the polymorphic norm are the howler monkeys (*Alouatta sp*). Like the catarrhine primates, howler monkeys have separate M and L cone opsin genes on their X-chromosome and so all of the monkeys in these species would be expected to have trichromatic color vision (Jacobs et al. 1996b). The gene duplication that led to this arrangement in howler monkeys seems to be unique to this lineage and, interestingly, it is not related to the duplication event that rendered all of the catarrhines trichromatic (Hunt et al. 1998).

The third major branch of the primate order is comprised of the more primitive strepsirrhines of Africa and southeast Asia (lemurs, bushbabies, etc). These primates show remarkable variation in their M/L pigment arrangements (Tan and Li 1999; Tan et al. 2005; Jacobs and Deegan II 2003). First, some of the diurnal species have M/L pigment polymorphisms similar in nature to those described for the platyrrhine monkeys. Although good tests of color vision are so far lacking, heterozygous females from these species would be predicted to have trichromatic color vision, whereas homozygous females and all males should be dichromatic. Second, other strepsirrhine species have pigment arrangements much like those of most non-primate mammals, i.e., an S cone pigment and a single M/L pigment. They presumably have dichromatic color vision. Finally, some nocturnal strepsirrhines are similar to *Aotus* monkeys (above) in that they have only a single type of M/L cone pigment and must therefore lack a color vision capacity.

In sum, it appears that a number of different evolutionary changes in the cone opsin genes have, in combination, shaped the nature of color vision in contemporary primates. These include 1) a loss of function of S-cone opsin genes in some platyrrhine and many strepsirrhine lineages, 2) the appearance of opsin gene polymorphisms in almost all platyrrhines and in a few strepsirrhine genera, and 3) the addition of a third opsin gene, an event triggered by an early gene duplication in catarrhines and by X-chromosome gene polymorphisms in some platyrrhine and strepsirrhine species. To complete the picture of variation in primate color vision, it should be noted that there are a number of polymorphisms of X-chromosome opsin genes that seem to be unique to humans. Such polymorphisms include not only the variations that underlie the classical types of defective color vision, characteristically seen only among males, but also a considerable variety of opsin gene variations that have been detected in both males and females. The latter seem to produce color vision changes that are at best quite subtle (Neitz and Neitz 2003). It appears that all of the opsin gene polymorphisms detected in humans are effectively absent in other catarrhine primates (Jacobs and Williams 2001; Onishi et al. 1999) and thus must have emerged during the 5- to 6-million-year history of our species. Just why humans are distinctive in that regard remains a mystery.

5 Evolution of SWS1 Cone Pigments

Cone pigments derived from genes in the SWS1 family (Fig. 2) fall into two spectral classes having respective peaks in the ultraviolet (UV) portion of the spectrum, λ_{max} of ~360–390 nm and at longer wavelengths (λ_{max} of ~395–435 nm). The former are usually referred to as UV pigments, the latter S pigments. The picture relating gene sequence changes and spectral tuning for the SWS1 pigments is more complex than it is for the LWS pigments. Comparative examinations of opsin gene sequences suggest that the UV pigments were ancestral (Hunt et al. 2004). In some lineages, these ancestral UV pigments were converted to S pigments and sometimes then reconverted back to UV pigments. The amino acid substitutions associated with tuning changes may involve as many as eight residues, with perhaps three of these being key to spectral shifting (Hunt et al. 2004).

Until fairly recently it was believed that, although the locations of their λ_{max} values may vary from species-to-species, all mammals had some form of an S pigment. But that idea has turned out to be untrue; in fact, some mammals have retained the ancestral UV sensitivity (Jacobs et al. 1991). Most mammals having UV pigments are rodents (among common examples, house mice and rats), although some rodents have acquired S pigments - for example, most members of the squirrel family - so the shift from UV to S pigments must have occurred in the more distal branches of rodent phylogeny. The former are nocturnal, the latter diurnal, so one might think that a change to a diurnal lifestyle is somehow linked to a shift from UV to S pigments. However, there are many mammals possessing S pigments that are classified as nocturnal, and at least some diurnal rodents have UV photopigments (Jacobs et al. 2003), so there seems to be no completely compulsive relationship between lifestyle and SWS1 pigment type. Further clouding this picture is the fact that it is not at all obvious how heightened sensitivity to the UV might particularly benefit a nocturnal animal.

An interesting twist to the story of mammalian SWS1 pigments is that there has been a loss of function of these opsin genes in some lineages. As noted above, this loss was first discovered during a search for S pigments in the retina of a nocturnal platyrrhine primate (*Aotus*). No S pigment could be found and, subsequently, it was discovered that its absence reflected the fact that the SWS1 genes in this species harbor fatal mutational changes that obviate the expression of any pigment. This loss seems to be species-wide, allowing the possibility that the loss of S pigment may be in some way adaptive. A similar loss of S-cone function was subsequently documented in some other primates, some rodents, and a few carnivores. It is particularly interesting that this loss of function of the S-cone opsin gene seems to be universal among many species of marine mammal, including all of the pinnipeds and cetaceans (Levenson and Dizon 2003; Levenson et al. 2006). At present we have no very satisfactory explanation of what caused a number of mammalian lineages to effectively abandon their S cones, and along with them the potential for a dimension of color vision and the additional visual sensitivity to short-wavelength radiations that S cones can provide.

6 Evolving Color Vision in the Laboratory

The comments above indicate that evolution of color vision in primates was triggered by opsin gene alterations that led to the appearance of new cone photopigments. What is unclear from this post hoc linkage is whether the addition of new pigments immediately provided a new dimension of color vision or whether subsequent nervous system changes were also required to capitalize on the altered pigment array. Genetic engineering can allow a direct test of these alternatives. To accomplish such a test, we recently studied a knock-in mouse that was designed to mimic the polymorphic photopigment arrangement characteristic of the platyrrhine monkeys (Jacobs et al. 2007). Like most non-primate mammals, the mouse has a single X-chromosome opsin gene, in this case encoding a pigment with a λ_{max} of 510 nm. This mouse opsin gene was replaced with one that normally encodes the human L-cone pigment ($\lambda_{max} = 556$ nm). Subsequent breeding yielded several possible X-chromosome gene arrays in these animals: males and homozygous females with either the 510-nm or 556-nm pigment and, by analogy to the platyrrhine monkeys, heterozygous females with both cone pigments. Behavioral tests revealed that these heterozygous females had acquired a novel color vision capacity, and this result implies that the mouse visual system is sufficiently plastic to make use of information provided by the presence of a new photopigment. This experiment thus provides support for the contention that the mere addition of a new cone pigment through opsin gene changes probably immediately added some new color vision to the visual repertoire of our primate ancestors.

7 The Utility of Color Vision

The mechanistic steps for adding new pigments or for altering pigments already present in the eye through opsin gene changes are now reasonably understood. For these changes to be maintained through successive generations, i.e., to evolve, additionally requires that they somehow increase fitness. That requirement raises the question of what advantages may accrue from alterations in color vision.

Without color vision an animal cannot reliably distinguish between wavelength and intensity changes in the photic environment. In essence, having color vision allows an animal to disambiguate variations in the intensities of lights from variations in the spectral energy distributions of lights. The extent of that disambiguation depends on the dimensionality of the color vision. For simplicity, consider only a single case, that of differences between dichromatic and trichromatic color vision. Details about the characteristics of these two forms of color vision have been accumulated in hundreds of different laboratory experiments. One fundamental aspect of the two arrangements is illustrated in Figure 4a, which shows wavelength discrimination functions for dichromatic and trichromatic monkeys. The stimuli in such experiments are monochromatic lights, and the advantage of adding a dimension of color vision in such circumstances is obvious: the trichromatic animal is able to reliably detect subtle differences among middle and long wavelength lights, none of which can be discerned by the dichromat. Adding a second middle- to long-wavelength pigment not only enhances color discrimination but also provides an expansion of the total spectral window of vision. That advantage can be seen in Figure 4b, which shows spectral sensitivity functions obtained from these same dichromatic and trichromatic monkeys. It is obvious that adding a second pigment allows the trichromat to achieve greatly enhanced sensitivity to the long wavelengths. Taken together, the results shown in Figure 4 reveal an important advantage of adding a pigment to the retinal complement: that change sets the stage for allowing animals to successfully discriminate among many more items in their visual environments. The quantitative extent of this advantage may be quite impressive; for example, some estimates suggest that the total number of wavelength and intensity combinations that an animal can discriminate is about one thousand times greater for the trichromat as compared to the dichromat (Neitz et al. 2001).

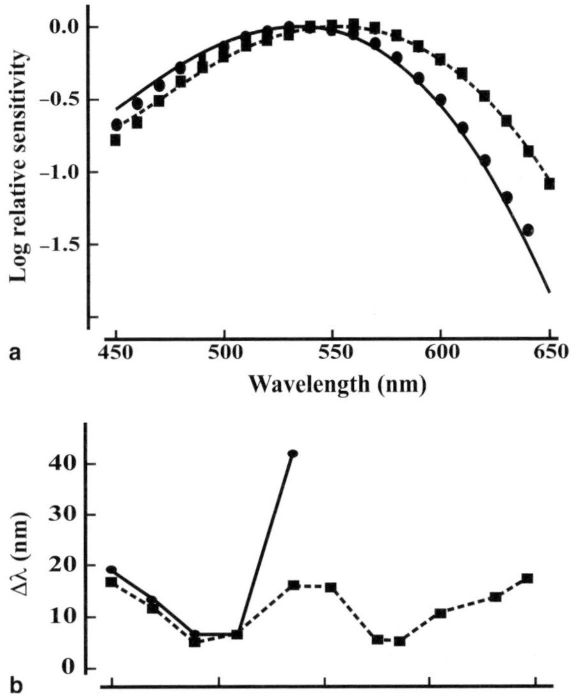

Fig. 4 Visual discrimination results obtained from behavioral tests of dichromatic (solid circles) and trichromatic (solid squares) platyrrhine monkeys. These animals had the photopigment arrangements illustrated in the bottom panel of Figure 3. **a** Spectral sensitivity functions obtained from electroretinogram measurements made on dichromatic and trichromatic animals. **b** Wavelength discrimination functions obtained from the same animals whose results are shown in panel **a**. These results are further described in the text

Although it is not hard to see potential advantages of trichromatic over dichromatic vision in laboratory studies, it is considerably more challenging to understand how such results may translate to natural environments. Over the course of their history, many primates have evolved trichromatic color vision and, consequently, these animals provide a natural resource for examining the utility of trichromatic color vision. To survive, animals from any given species must find sustenance, successfully avoid dangers, find mates, and reproduce. How might color vision aid in this complex of behaviors? The focus has often fallen on the first of these tasks and, since many primates are fruit eaters, it is not surprising that one persistent suggestion about the evolution of primate color vision is that it was closely linked to the evolution of fruit coloration (e.g., Allen 1879; Polyak 1957). Attempts to evaluate this possibility through direct experimentation are more modern, being principally fostered by the recent widespread availability of devices that allow accurate measurements of light in the wild, which can then be efficiently manipulated, and by the growing understanding of the neural mechanisms underlying primate color vision. Starting with light measurements of fruits, foliage, and natural illuminants, many investigators have used computational models incorporating photopigment spectra and neural networks to evaluate how trichromats and dichromats might fare in discriminations of these natural objects. There is widespread agreement among such studies that trichromats should be superior to dichromats in detecting red, green and yellow fruits against foliage backgrounds (Osorio and Vorobyev 1996; Regan et al. 2001; Parraga et al. 2002), but there is little agreement as to whether these particular types of discriminations were in fact ones critical for the evolution of primate trichromacy. Partly this uncertainty is because both across and within-species the diets of primates can be highly variable, involving not only fruits but also foliage, gums, insects, and other items. Given this variability, one alternative that has been proposed is that color may provide a particularly useful cue for the detection of the young leaves that form a common dietary choice for some species and that often (to the human eye) exhibit a characteristic reddish flush in tropical climes (Dominy and Lucas 2001). Model calculations show that trichromatic color vision is in fact also superior to dichromatic color vision in detection tasks of this sort (Dominy and Lucas 2001; Sumner and Mollon 2000). Indeed, it can be shown that trichromacy is also better suited than dichromacy for another potentially important visual task: detecting variations in skin (and, presumably, pelage) colors (Changizi et al. 2006). In sum, this body of work agrees that trichromacy should be superior to dichromacy in the performance of a number of tasks that are important for primate success, but whether one or the other of these classes of natural discriminations involving color would have been uniquely critical for its evolution remains very much uncertain. Since primates are visual generalists, one possibility is that all of these types of discriminations may have been important for survival and thus supported the evolution of trichromacy.

Computational studies such as those just described can provide useful insights into the potential utility of different forms of color vision, but by themselves they cannot be definitive. One alternative approach is to determine through direct observational studies how monkeys do use color in the conduct of their normal affairs, e.g., in the harvesting of palatable foods. For such tasks, the polymorphic platyrrhine monkeys offer great advantages, since in any group of conspecific animals all of them must share common goals, but individual monkeys have to achieve these goals by employing differing color vision capacities. Although simple in concept, in practice such studies offer many challenges, as platyrrhine monkeys are mostly small, fast moving, and visually elusive in the wild, so assessing just what they are using their eyes for at any moment is difficult. There have been a number of recent field studies of this type, but perhaps due to such difficulties they do not seem to have provided much in the way of useful insights into the problem, at least not to date (Jacobs 2007).

One recent and interesting variant on these observational studies has involved the examination of captive tamarin monkeys in a semi-natural feeding setting (Smith et al. 2003). The objects that the monkeys were asked to harvest were small boxes containing palatable foods that were marked with colored papers designed to match the reflectance properties of unripe, partially ripe, and fully ripe fruits of a type that forms an important part of the diet of these animals in the wild. These fruits were displayed so that they had to be detected in a similarly realistic "leafy" background. Trichromatic animals were found to be quicker than the dichromatic animals to associate fruit color with the baited boxes and faster to acquire the "fully ripe fruits" that contained the largest amount of reward. The experiment thus directly supports the predictions from the computational studies that trichromats have a foraging advantage when targets are signaled by realistic color cues. There have been a number of other studies carried out in a similar fashion and the usual conclusion is that trichromatic animals hold a foraging advantage over dichromats, although the differences between animals having the two forms of color vision are often rather small, making it hard to know how these results might translate to real-world settings.

As described, the polymorphic platyrrhine monkeys vary both between dimensional categories and within them, yielding multiple types of trichromatic and dichromatic color vision. Simply lumping all these animals into two categories, as has been done in the discussion above, is apt to miss much of the nuance of color vision variations. For example, the trichromatic subvarieties of platyrrhine monkeys differ in the spectral separations of their M and L pigments (Fig. 3b). What influence on color vision might such differences hold? Although we do not yet have a complete answer to that question, there are hints from both computational studies (Osorio et al. 2004) and from simulation studies involving human subjects (Rowe and Jacobs 2004, 2006) that the trichromatic subtypes that feature the largest spectral separation of their M and L pigments are generally advantaged in the performance of red/green discriminations. The latter experiments also point to the importance of viewing time and overall light levels in assessing the relative virtues of different trichromatic and dichromatic forms of color vision.

In sum, although color specialists have made a start toward understanding the functional utility of color vision, and thus ultimately revealing why particular forms of color vision may have been evolutionarily favored, much still remains to be done. One major issue is that the problems foraging monkeys face are more complex than have been effectively studied to date. In trying to move toward understanding these more naturalistic demands, one possible suggestion is that the challenges that behaving animals must meet may be similar to what has been extensively studied in humans in so-called visual search tasks. In such tests the visual field contains an array of items that are marked with multiple cues (color, size, shape, etc), something similar to what one might find in rich natural environments. Several of the cue dimensions in the search are irrelevant to the detection task, and they thus can serve as distractors. When color cues are specifically studied in such contexts, it appears that human color defectives are generally less efficient in carrying out target searches in a cluttered environment than are individuals with normal color vision (Cole et al. 2004). To better understand the utility of the different color vision arrangements that have emerged among the various primates it may ultimately prove profitable to apply similar sorts of tests to them.

References

Allen G (1879) The colour-sense: its origin and development. Boston: Houghton, Osgood, & Company.

Arrese CA, Hart NS, Thomas N, Beazley LD, Shand J (2002) Trichromacy in Australian marsupials. Curr Biol 12:657–660.

Asenjo AB, Rim J, Oprian DD (1994) Molecular determinants of human red/green color discrimination. Neuron 12:1131–1138.

Bowmaker JK, Astell S, Hurst DM, Mollon JD (1991) Photosensitive and photostable pigments in the retinae of Old World monkeys. J Exp Biol 156:1–19.

Brown PK, Wald G (1963) Visual pigments in human and monkey retinas. Nature 200:37–43.

Byrne A, Hilbert DR (2003) Color realism and color science. Behav Brain Sci 26:3–64.

Changizi MA, Zhang Q, Shimojo S (2006) Bare skin, blood and the evolution of primate colour vision. Biol Lett 2:217–221.

Cole BL, Maddocks JD, Sharpe K (2004) Visual search and the conspicuity of coloured targets for colour vision normal and colour vision deficient observers. Clin ExpOptom 87:294–304.

Collin SP, Trezise AEO (2004) The origins of colour vision in vertebrates. Clin Exp Optom 87:217–233.

Dalton J (1924) Letter to his cousin, February 1794. Memoirs and Proceedings of the Manchester Literary and Philosophical Society 68:113–117.

Davies WL, Carvalho LS, Cowing JA, Beazley LD, Hunt DM, Arrese C (2007) Visual pigments of the platypus: A novel route to mammalian colour vision. Curr Biol 17:R161–R163.

Dominy NJ, Lucas PW (2001) Ecological importance of trichromatic colour vision to primates. Nature 410:363–365.

Hardin CL (1988) Color for philosophers. Indianapolis: Hackett Publishing Company.

Hilbert DR (1987) Color and color perception: A study in anthropocentric realism. Stanford CSLI/Stanford.

Hisatomi O, Tokunaga F (2002) Molecular evolution of proteins involved in vertebrate phototransduction. Comp Biochem Physiol B 133:509–522.

Hunt DM, Dulai KS, Bowmaker JS, Mollon JD (1995) The chemistry of John Dalton's color blindness. Science 267:984–988.

Hunt DM, Dulai KS, Cowing JA, Juillot C, Mollon JD, Bowmaker JK, Li W-H, Hewett-Emmett D (1998) Molecular evolution of trichromacy in primates. Vision Res 38:3299–3306.

Hunt DM, Cowing JA, Wilkie SE, Parry JWL, Poopalasundaram S, Bowmaker JK (2004) Divergent mechanisms for the tuning of shortwave sensitive visual pigments in vertebrates. Photochem Photobiol Sci 3:713–720.

Jacobs GH (1984) Within-species variations in visual capacity among squirrel monkeys (*Saimiri sciureus*): Color vision. Vision Res 24:1267–1277.

Jacobs GH (1993) The distribution and nature of colour vision among the mammals. BiolRev 68:413–471.

Jacobs GH (2007) New World monkeys and color. Intl J Primatol 28:729–759.

Jacobs GH, Deegan II JF (1999) Uniformity of colour vision in Old World monkeys. Proc Royal Soc London B 266:2023–2028.

Jacobs GH, Deegan II JF (2003) Diurnality and cone pigment polymorphism in strepsirrhines: Examination of linkage in *Lemur catta*. Am J Phys Anthropol 122:66–72.

Jacobs GH, Rowe MP (2004) Evolution of vertebrate colour vision. Clin Exp Optom 87:206–216.

Jacobs GH, Williams GA (2001) The prevalence of defective color vision in Old World monkeys and apes. Color Res Appli 26:S123–S127.

Jacobs GH, Neitz J, Crognale M (1987) Color vision polymorphism and its photopigment basis in a callitrichid monkey (*Saguinus fuscicollis*). Vision Res 27:2089–2100.

Jacobs GH, Neitz J, Deegan II JF (1991) Retinal receptors in rodents maximally sensitive to ultraviolet light. Nature 353:655–656.

Jacobs GH, Deegan II JF, Neitz JA, Crognale MA, Neitz M (1993) Photopigments and color vision in the nocturnal monkey, *Aotus*. Vision Res 33:1773–1783.
Jacobs GH, Neitz M, Neitz J (1996a) Mutations in S-cone pigment genes and the absence of colour vision in two species of nocturnal primate. Proc Royal Soc London B 263:705–710.
Jacobs GH, Neitz M, Deegan JF, Neitz J (1996b) Trichromatic colour vision in New World monkeys. Nature 382:156–158.
Jacobs GH, Calderone JB, Fenwick JA, Krogh K, Williams GA (2003) Visual adaptations in a diurnal rodent, *Octodon degus*. J Comp Physiol A 189:347–361.
Jacobs GH, Williams GA, Cahill H, Nathans J (2007) Emergence of novel color vision in mice engineered to exress a human cone photopigment. Science 315:1723–1725.
Kelber A, Vorobyev M, Osorio D (2003) Animal colour vision–behavioural tests and physiological concepts. Biol Rev 78:81–118.
Lashley KS (1934) Introduction. In: Schiller CH (ed) Instinctive behavior: The development of a modern concept (New York: International Universities Press, Inc., pp ix–xii.
Levenson DH, Dizon A (2003) Genetic evidence for the ancestral loss of SWS cone pigments in mysticetee and odontocete cetaceans. Proc Royal Soc London B 270:673–679.
Levenson DH, Ponganis PJ, Crognale MA, Deegan II JF, Dizon A, Jacobs GH (2006) Visual pigments of marine carnivores: Pinnipeds, polar bear, and sea otter. J Comp Physiol A 192:833–843.
Lubbock J (1882) Ants, bees and wasps: a record of observations on the habits of the social hymenoptera. London: New edition, based on the 17th, 1929. Keegan Paul, Trench and Trubner and Co.
Merbs SL, Nathans J (1993) Role of hydroxyl-bearing amino acids in differentially tuning the absorption spectra of the human red and green cone pigments. Photochem Photobiol 58:706–710.
Nathans J, Thomas D, Hogness DS (1986) Molecular genetics of human color vision: the genes encoding blue, green and red pigments. Science 232:193–202.
Neitz J, Carroll J, Neitz M (2001) Color vision: Almost reason enough for having eyes. Optics Photonics News 12:26–33.
Neitz M, Neitz J (2003) Molecular genetics of human color vision and color vision defects. In: Chalupa LM, Werner JS (eds) The visual neurosciences.Cambridge: MIT Press, pp 974–988.
Neitz M, Neitz J, Jacobs GH (1991) Spectral tuning of pigments underlying red-green color vision. Science 252:971–974.
Newton I (1730) Optics. Dover, 1952 Edition.
Onishi A, Koike, S, Ida M, Imai H, Schichida Y, Osamu T, Hanazawa A, Konatsu H, Mikami A, Goto S, Suryobroto B, Kitahara K, Yamamori T (1999) Dichromatism in macaque monkeys. Nature 402:139–140.
Osorio D, Vorobyev M (1996) Colour vision as an adaptation to frugivory in primates. Proc Royal Soc London B 263:593–599.
Osorio D, Smith AC, Vorobyev M, Buchanan-Smith HM (2004) Detection of fruit and the selection of primate visual pigments for color vision. Am Natural 164:696–708.
Parraga CA, Troscianko T, Tolhurst DJ (2002) Spatiochromatic properties of natural images and human vision. Curr Biol 12:483–587.
Polyak S (1957) The vertebrate visual system. Chicago: University of Chicago Press.
Regan BC, Julliot C, Simmen B, Vienot F, Charles-Dominique P, Mollon JD (2001) Fruits, foliage and the evolution of primate colour vision. Phil Trans Royal Soc London B 356:229–283.
Rowe MP, Jacobs GH (2004) Cone pigment polymorphisms in New World monkeys: Are all pigments created equal? Visual Neurosci 21:217–222.
Rowe MP, Jacobs GH (2006) Naturalistic color discrimination in polymorphic platyrrhine monkeys: Effects of stimulus luminance and duration examined with functional substitution. Visual Neurosci 24:17–23.
Smith AC, Buchanan-Smith HM, Surridge AK, Osorio D, Mundy NI (2003) The effect of colour vision status on the detection and selection of fruits by tamarins (Saguinus spp.). J Exp Biol 206:3159–3165.
Solomon SG, Lennie P (2007) The machinery of colour vision. Nature Neurosci Rev 8:276–286.

Sumner P, Mollon JD (2000) Catarrhine photopigments are optimized for detecting targets against a foliage background. J Exp Biol 203:1963–1986.

Tan Y, Li W-H (1999) Trichromatic vision in prosimians. Nature 402:36.

Tan Y, Yoder AD, Yamashita N, Li W-H (2005) Evidence from opsin genes rejects nocturnality in ancestral primates. Proc Natl Acad Sci USA 41:14712–14716.

Tovee MJ, Bowmaker JK, Mollon JD (1992) The relationship between cone pigments and behavioural sensitivity in a New World monkey (*Callithrix jacchus jacchus*). Vision Res 32:867–878.

von Uexküll J (1934) A stroll through the worlds of animals and men. In: Schiller CH (ed) Instinctive behavior: The development of a modern concept. (New York: International Universities Press, Inc., pp. 5–80.

Yokoyama S, Radlwimmer FB (1998) The "five-sites" rule and the evolution of red and green color vision in mammals. Mol Biol Evol 15:560–567.

The Evolution of Social Categories

Robert M. Seyfarth and Dorothy L. Cheney

Abstract A widely accepted view holds that human infants' knowledge of the world develops not from a single, general-purpose ability to reason about objects and events but instead from several "core" systems of knowledge, each specialized for representing and reasoning about entities of different kinds. One functionally specialized system concerns knowledge about animate creatures. If knowledge about animate beings – particularly conspecifics – constitutes a core system of cognition in human infants, it may be possible to explain its evolution by 1) identifying a core system of social knowledge in nonhuman primates and other animals, 2) demonstrating that this knowledge constitutes a functionally specialized system and 3) demonstrating that knowledge of other individuals' identities, category membership, and motives is adaptive, contributing to the reproductive success of individuals. Here we test these predictions using data from wild baboons. Results indicate that baboons classify each other along a number of dimensions simultaneously: as individuals, members of matrilineal kin groups, holders of specific dominance ranks, and participants in more transient social relationships. Their formation of social categories is based on certain expectations and assumptions of causality. Baboon social knowledge, moreover, is adaptive because it allows individuals to form relationships that reduce stress and increase reproductive success.

1 Introduction

A widely accepted view holds that human infants' knowledge of the world develops not from a single, general purpose ability to reason about objects and events but instead from several "core" systems of knowledge, each specialized for representing and reasoning about entities of different kinds (Hirschfeld and Gelman 1994; Carey and Spelke 1996). This view of mental function as a collection of domain-specific

R.M. Seyfarth and D.L. Cheney
University of Pennsylvania, Departments of Biology & Psychology,
3720 Walnut St. Room D7, Philadelphia, PA 19104 USA
e-mail: seyfarth@psych.upenn.edu, cheney@sas.upenn.edu

A. Berthoz and Y. Christen (eds.), *Neurobiology of "Umwelt": How Living Beings Perceive the World,* Research and Perspectives in Neurosciences,

"modules" (Fodor 1983) or functional specializations (Barrett and Kurzban 2006) comes in many different forms. The more extreme proponents of the modular view (e.g., Cosmides and Tooby 1994) write as if general learning processes like classical and instrumental conditioning, which obey similar rules across many different domains, have no place in the explanation of human behavior. Others willingly accept that general modes of conditioning can operate within a domain, and experience of a general sort can in some cases lead to domain-specific expertise (Gallistel et al. 1991; Shettleworth 1998). Studies of language acquisition in children offer an excellent example of research in which there is an active debate about the relative importance of generalized and specialized learning mechanisms (compare Jackendoff 1994 and Pinker 1994 with Tomasello 2003). Regardless of their stance on the relative importance of domain-specific and domain-general processes, however, all investigators agree that it can be difficult to establish precisely the borders of a cognitive "domain."

Theories of domain-specific knowledge are evolutionary theories: they assume that our cognitive "architecture" takes the form that it does because, during the evolutionary history of humans, other primates, and perhaps other birds and mammals, one style of perception and cognition has given individuals an advantage in survival or reproduction over others who perceive the world in a different way (Barkow et al. 1992). Supporting this view, domain-specific hypotheses receive considerable support from the ethological literature. For example, in her review of animal learning and behavior, Shettleworth concludes: "Our review of how animals perceive, ... learn, ... recognize, and categorize, ... makes clear that the animal mind contains a variety of adaptively specialized cognitive modules" (1998, p. 566). As in studies of infant development, the many candidates for separate cognitive modules in animals "are distinguished from one another by domain, representational content, and possible distinctive rules of operation." This is not to say, however, that the animal mind is exclusively modular, because domain-specific processing can coexist with associative processes that obey the same general rules across many sensory modalities and behavioral contexts. As Shettleworth puts it, "associative learning has a variety of adaptive uses" (1998, p. 568).

Knowledge of animate creatures, or "agents." is often cited as one of the human infant's core systems of knowledge. Studies supporting this view have examined infants' and children's knowledge of animate and inanimate objects (Gelman et al. 1983), their early recognition of faces and voices (e.g., Kleiner 1993; Vouloumanos and Werker 2007), biological classification (Keil 1989), and attribution to others of mental states like knowledge, motives, and beliefs (e.g., Leslie 1987; Wellman 1990), as well as adults' organization of individuals into social categories (e.g., Atran 1990; Hirschfeld 1994). Taken together, these studies suggest that children are born with an innate, domain-specific sensitivity to animate as opposed to inanimate objects, and within the former group to humans as opposed to other creatures. From these early perceptual and cognitive specializations, infants and children rapidly develop concepts and theories about people as unique individuals who can be classified into certain social categories or groups and whose behavior is closely linked to their mental states.

If knowledge about animate beings – particularly conspecifics – constitutes a core systems of cognition in human infants, it may be possible to explain its evolution by 1) identifying a core system of social knowledge and classification in nonhuman primates and other animals, 2) demonstrating that this knowledge involves domain-specific, functionally specialized processing, and 3) demonstrating that knowledge of others' identities, category membership, and motives is adaptive, contributing to individuals' reproductive success. In this paper we begin with a brief review of the brain areas in primates that appear to be specialized for processing social stimuli. We then review what is known about the social knowledge of one primate species, the baboon (*Papio cynocephalus ursinus*), and consider its adaptive value. We then discuss whether baboon social knowledge constitutes an innate, domain-specific specialization and whether it involves the construction of concepts or theories. We conclude with some speculations about the primacy of social categories during the course of primate and human evolution.

2 Neural Correlates of Social Knowledge

Monkeys have "face cells" in the temporal cortex that respond at least twice as vigorously to faces or components of faces (like eyes or mouths) than to other complex visual stimuli (Tsao et al. 2003, 2006). Face cells are surprisingly specialized. Those in the inferior temporal cortex (IT) seem most important for processing facial identity, whereas those in the superior temporal sulcus (STS) seem most important for processing facial expressions. IT and STS are extensively interconnected and probably share face-specific information (Weiss et al. 2002; Ghazanfar and Santos 2004). Face cells in STS respond not only to facial expressions but also to the direction of an individual's head orientation and gaze. Their response is greatest when head orientation and gaze direction are congruent, less strong when they are incongruent (Emery and Perrett 2000; Jellema et al. 2000; Perrett et al. 1992; Eifuku et al. 2004). The STS of rhesus macaques also includes neurons that fire when the monkey observes an individual walking, turning his head, bending, or extending his arm (Perrett et al. 1990).

Particularly intriguing are the "mirror-neurons" in the inferior parietal lobule that show elevated activity both when the subject monkey executes a specific grasping action and when the monkey observes a human or other monkey execute a more or less similar grasp (Rizzolatti and Craighero 2004). Neurons that code for specific acts, such as grasping, also seem to be context-dependent. Some neurons respond more when the monkey grasps a piece of food to eat it than when he grasps the same food to place it into a container. This same context-dependence is preserved when the monkey observes *another* individual perform these actions. Significantly, many neurons begin to fire *before* the other individual actually performs a specific action, that is, before he grasps-to-eat as opposed to grasps-to-place. Thus, it seems possible that these neurons encode not only the specific motor act but also the actor's intentions (Fogassi et al. 2005; see also Nakahara and Miyashita 2005; Rizzolatti and Craighero 2004; Rizzolatti and Buccino 2005).

Finally, monkeys – like humans – process their own species' vocalizations in ways that are measurably different from the way they process other auditory stimuli. Like humans, monkeys display a left-brain, right-ear advantage when processing their own species' vocalizations, but not when processing other sounds (Petersen et al. 1978; Hefner and Hefner 1984; Weiss et al. 2002; Poremba et al. 2004; but contrast Hauser and Andersson 1994 with Teufel et al. 2007). The processing of conspecific vocal signals also has a strong bi-modal component. Monkeys recognize the correspondence between facial and vocal expressions (Ghazanfar and Logothetis 2003), and when rhesus macaques hear another monkey's calls, they exhibit neural activity not only in areas associated with auditory processing but also in higher-order visual areas (Gil da Costa et al. 2004).

Specialized cells and unique brain mechanisms for dealing with socially relevant stimuli are not definitive proof of domain-specific social intelligence in primates. But they are just what we would expect to find if natural selection had acted with particular force to favor individuals skilled in solving social problems (for further discussion, see Cheney and Seyfarth 2007).

3 Social Knowledge in Baboons

Baboons are Old World monkeys that shared a common ancestor with humans roughly 30 million years ago (Steiper et al. 2004). They live throughout the savannah woodlands of Africa in groups of 50 to 150 individuals. Although most males emigrate to other groups as young adults, females remain in their natal groups throughout their lives, maintaining close social bonds with their matrilineal kin (Silk et al. 1999, 2006a, b). Females can be ranked in a stable, linear dominance hierarchy that determines priority of access to resources. Daughters acquire ranks similar to those of their mothers. The stable core of a baboon group is therefore a hierarchy of matrilines, in which all members of one matriline (for example, matriline B) outrank or are outranked by all members of another (for example, matrilines C and A, respectively). Ranks within matrilines are as stable as those between matrilines: for example, $A1 > A2 > A3 > B1 > B2 > C1$, where letters are used to denote matrilineal kin groups and numbers denote the different individuals within them (Cheney and Seyfarth 2007).

Baboon vocalizations, like those of many other primates, are individually distinctive (e.g., Owren et al. 1997; Rendall 2003), and playback experiments have shown that listeners recognize the voices of others as the calls of specific individuals (reviewed in Cheney and Seyfarth 2007). The baboon vocal repertoire contains a number of acoustically graded signals, each of which is given in predictable contexts. Because calls are individually distinctive and each call type is predictably linked to a particular social context, baboon listeners can potentially acquire quite specific information from the calls that they hear. Experiments that test what information actually is obtained are thus, in effect, tests of social cognition.

Throughout the day, baboons hear other group members giving vocalizations to each other. Some interactions involve aggressive competition, for example, when a higher-ranking animal gives a series of threat-grunts to a lower-ranking animal and the latter screams. Threat-grunts are aggressive vocalizations given by higher-ranking to lower-ranking individuals, whereas screams are submissive signals, given primarily by lower- to higher-ranking individuals. A threat-grunt-scream sequence, therefore, potentially provides information not only about the identities of the opponents involved but also about who is threatening whom. Baboons are sensitive to both types of information. In playback experiments, listeners respond with apparent surprise to sequences of calls that appear to violate the existing dominance hierarchy. Whereas they show little response upon hearing the sequence "B2 threat-grunts and C3 screams," they respond strongly – by looking toward the source of the call – when they hear "C3 threat-grunts and B2 screams" (Cheney et al. 1995a). Between-family rank reversals (C3 threat-grunts and B2 screams) elicit a stronger violation of expectation response than do within-family rank reversals (C3 threat grunts and C1 screams; Bergman et al. 2003).

A baboon who ignores the sequence "B2 threat-grunts and C3 screams" but responds strongly when she hears "C3 threat-grunts and B2 screams" reveals, by her responses, that she recognizes the identities of both participants, their relative ranks, and their family membership. Baboons who react more strongly to call sequences that mimic a between-family rank than to those that mimic a within-family rank reversal act as if they classify individuals simultaneously according to both rank and kinship (Bergman et al. 2003). In all of these cases, listeners act as if they assume that the threat-grunt and scream have occurred together not by chance, but because one vocalization caused the other to occur. Without this assumption of causality there would be no violation of expectation when B2's scream and C3's threat-grunt occurred together.

Baboons' ability to deduce a social narrative from a sequence of sounds reveals a rich cognitive system in which listeners extract a large number of complex, nuanced messages from a relatively small, finite number of signals. A baboon who understands that "B2 threat-grunts and C3 screams" is different from "C3 threat-grunts and B2 screams" can make the same judgment for all possible pairs of group members as well as any new individuals who may join (Cheney and Seyfarth 2007).

In addition to making judgments based on social causation, rank, and kinship, baboons appear to recognize other individuals' intentions and motives. Baboon groups are noisy, tumultuous societies, and a baboon would not be able to feed, rest, or engage in social interactions if she responded to every call as if it were directed at her. In fact, baboons seem to use a variety of behavioral cues, including gaze direction, learned contingencies, and the memory of recent interactions with specific individuals when making inferences about the target of a vocalization. For example, when a female hears a recent opponent's threat-grunts soon after fighting with her, she responds as if she assumes that the threat-grunt is directed at her, and she avoids the signaler. However, when she hears the same female's threat-grunts soon after grooming with her, she acts as if the calls are directed at someone else and ignores the calls (Engh et al. 2006a).

The attribution of motives to others is perhaps most evident in the case of "reconciliatory" vocalizations. Like many other group-living animals, baboons incur both costs and benefits from joining a group. In an apparent attempt to minimize the disruptive effects of within-group competition, many primates "reconcile" with one another by coming together, touching, hugging, or grooming after aggression (Cheney et al. 1995b). In baboons, reconciliation among females occurs after roughly 10% of all fights, and typically occurs when the dominant animal grunts to the subordinate (Cheney et al. 1995b; Silk et al. 1996). Playback experiments have shown that, even in the absence of any other behavior, grunts alone function to restore former opponents' behavior to baseline levels. When a subordinate female hears her opponent's grunt soon after a fight, she approaches her opponent and tolerates her opponent's approaches at a rate that is even higher than baseline rates (Cheney and Seyfarth 1997). In contrast, hearing the grunt of another, previously uninvolved high-ranking female unrelated to her opponent has no effect on the subordinate's behavior.

In some cases, the behavior of subordinates after aggression seems to involve more complex and indirect causal reasoning about both other animals' motives and their kinship bonds. Playback experiments have shown that baboons will accept the "reconciliatory" grunt by a close relative of a recent opponent as a proxy for direct reconciliation by the opponent herself (Wittig et al. 2007). If individual D1 has been threatened by individual A1 and then hears a grunt from A2, in the hour that follows she is more likely to approach, and more likely to tolerate the approaches of, A1 and A2 than if she had heard no grunt or a grunt from another high-ranking individual unrelated to the A matriline. Intriguingly, D1's behavior toward other members of the A matriline does not change. Subjects in these experiments act as if they recognize that a grunt from a particular female is causally related to a previous fight, but only if the caller is a close relative of her former opponent.

4 The Function of Social Cognition

Baboons live in a society where reproductive success depends upon social skills. Among female baboons living in Kenya, longevity and infant survival are the best predictors of reproductive success, and the best predictor of infant survival is the extent of a female's social integration (Silk et al. 2003). Females are strongly motivated to form close social bonds (as measured by frequent proximity and grooming) with others, particularly their matrilineal kin. When a mother dies, females respond by strengthening their bonds with other matrilineal kin. When few or no matrilineal kin are available, they form bonds with paternal sisters and/or unrelated individuals (Silk et al. 2006a).

In our study, female longevity and infant survival are also the primary determinants of a female's lifetime reproductive success (Cheney et al. 2004), and females experience the greatest stress from predation and infanticide – the two events that exert the greatest effect on their own and their infants' survival and reproduction. When her infant is threatened with infanticide, a female can alleviate stress by forming a temporary pair bond, or "friendship" with an adult male; when a close companion is killed by a predator, she can alleviate stress by broadening and extending her network of bonds with other females (Beehner et al. 2005; Engh et al. 2006b, c). During a "calm" period without infanticide or predation, females whose grooming networks focused on a few preferred partners had lower baseline glucocorticoid levels than did females whose grooming was spread more widely among others (Crockford et al. 2008). When the same females were subsequently challenged by the threat of infanticide, individuals with the most focused grooming networks and the lowest baseline glucocorticoid levels showed the smallest increase. All females decreased their grooming diversity when challenged, and those who showed the greatest decrease in grooming diversity had the smallest increase in glucocorticoid levels (Wittig et al. 2008).

In sum, a female baboon's skill in forming and maintaining social relationships affects her ability to overcome stressful situations and, ultimately, her reproductive success. Under these conditions, an individual who seeks to form and maintain those social relationships that return the greatest benefit must know as much as possible about other animals' relations, that is, she must have a sophisticated understanding of the individuals in her group, their long-term associations, short-term bonds, and the motivation that underlies them. Natural selection has thus favored the evolution of skills in identifying and classifying conspecifics because these skills are essential to survival and reproduction (Cheney and Seyfarth 2007).

5 Defining the Domain

The baboons' social domain consists of other baboons. We draw this conclusion because individuals single out other baboons and respond to them in ways that differ from the way they respond to, for example, other animals, rocks, trees, or clouds. Baboons recognize others as individuals and attribute to their conspecifics properties that they do not attribute to other organisms. They also perform operations (or computations) on individual baboons that they do not perform on other species.

Baboons seem to recognize that many animals have the capacity for self-generated motion and are therefore different from inanimate objects. When walking among impala, elephants, and warthogs, for instance, they appear to anticipate these animals' behavior and avoid them (particularly elephants) if they approach. Baboons thus have a kind of ontological category *animal* that excludes immobile objects. Within this category, they may associate different vocalizations with different species (see Seyfarth and Cheney 1990; Hauser and Wrangham 1990; and Zuberbuhler 2002a, b for evidence of such recognition in various primates), but there is as yet no evidence that they make finer discriminations, for example by recognizing different individuals within a species.

In contrast, baboons respond to each other as *individuals*: they link specific faces and bodies with particular voices to form a percept that is independent of modality. When a baboon hears a grunt from Sylvia she forms an expectation that, if she looks toward the sound, Sylvia is whom she will see. Baboons further assume that each individual has *motives*; they assume, in other words, that under certain circumstances the individual will behave toward specific others in particular ways. If a female hears a juvenile's distress scream, she looks toward the juvenile's mother (Cheney and Seyfarth 1999); if she receives aggression from Sylvia and soon thereafter hear Sylvia's threat grunt, she assumes that the threat grunt is directed at her (Engh et al. 2006a).

Baboons also classify other baboons according to their place in the group. Such classification occurs along several different dimensions. An individual occupies a specific dominance rank, for example, and is grouped with other individuals in a particular matriline. These classifications have several interesting properties. First, the inclination to rank others occurs not just among females, whose ranks are stable for years, but also among males, whose ranks change often (Kitchen et al. 2005).

Second, classifications are not based on physical appearance. Higher-ranking animals are not older, larger, or different in any obvious way from lower-ranking animals, and as far as we can tell members of the same matriline do not look alike. From its first day of life, an infant baboon is associated with its matrilineal kin by others, even though, as an infant, it looks very different from these animals.

Third, classification is genuinely "nested:" although the members of a matriline are associated together, within this group they retain their individual identities. Kin-biased reconciliation provides a good example. If individual D1 receives aggression from A1 and then hears a grunt from an individual unrelated to A1, the grunt has no affect on D1's behavior. However, if the grunt comes from a member of A1's matriline (say, A2), it acts as a reconciliatory signal, changing D1's behavior toward A1 and A2 but not toward any other members of the A matriline (Wittig et al. 2007). D1 behaves as if she recognizes that members of the A family "go together" but nonetheless remain distinct individuals.

Fourth, classifications persist despite births, deaths, and other demographic changes. This property suggests that baboon social knowledge is to some degree abstract, because the social categories of rank and matrilineal kinship persist despite changes in the individuals who comprise them.

This level of recognition and classification may be unique to conspecifics. Although no experiments have been conducted, it seems reasonable to assume that baboons do not, for instance, distinguish one rock from another and place all rocks into the same general category, or distinguish among different individual giraffes. Plants that provide food may constitute an exception. Experiments by Menzel (1991), for example, suggest that Japanese macaques distinguish different species of trees and associate, for each species, its own specific leaves, fruit, and location.

The privileged status of one's own species is not fixed and immutable: extensive experience with another species may overcome an individual's tendency to treat the members of her own species differently from all others. In a classic experiment, Humphrey (1974) tested rhesus macaques' interest in pictures of conspecifics and pictures of pigs. Without any training, subjects treated pictures of individual monkeys as different individuals but made no such distinction among pictures of individual pigs. However, after living for several weeks in a room where the wall was covered with pictures of different pigs, the same subjects responded to pictures of pigs in the same way that they responded to pictures of conspecifics – in both cases, distinguishing one individual from another. Hoesch (1961) describes a female baboon, Ahla, who was raised among goats on a farm in Namibia. As an adult, Ahla recognized individual goats by both their physical appearance and their voices. She also knew which kid was associated with each adult female. Ahla learned, in other words, to recognize in goats the same social categories that she would normally have recognized in baboons (Cheney and Seyfarth 2007).

Nonetheless, it seems unlikely that under normal circumstances baboons recognize and classify the members of other species in the same way that they recognize and classify other baboons. This is not surprising, because it is difficult to imagine how a baboon would benefit from recognizing, for example, different individual impala, or why natural selection would have favored this level of classification outside the social domain. If this conclusion is correct, then the borders of the baboons' social domain are clearly defined; for them, the "frame problem" is easily solved.

6 Is Primate Social Cognition Innate?

Thus far, all of our studies have examined adult subjects. As a result, we do not yet know whether the components of social intelligence – the recognition of others as individuals, the grouping of individuals according to rank and matrilineal kinship, and the "tracking" of transient relationships – constitute an innate property of the baboon's mind or are acquired only after extensive experience, through more general processes of perception, association, and memory.

Of course, even proponents of the most strongly innatist view must accept that some learning underlies baboons' social knowledge, because an infant baboon must learn the characteristics of the particular individuals that comprise her group. The question then becomes: is the infant baboon's mind a complete *tabula rasa*, ready to accept any social arrangement to which it might be exposed – monogamy, polygyny, with or without hierarchies or kin groups – or are infants born with an innate predisposition to recognize and classify individuals in some ways rather than others?

One hypothesis argues that baboons have an entirely unconstrained mind and that general learning mechanisms (classical and instrumental conditioning) are entirely sufficient to explain the development of their social knowledge, particularly if one accepts that such mechanisms can lead to rich, sophisticated representations of the elements that make up an animal's environment (Dickinson 1980; Rescorla 1988). Baboons are born into a social world that is filled with statistical regularities: when any two individuals come together, they behave in a predictable way toward each other. All a young baboon has to do is pay attention and remember. As they grow older and observe other animals' behavior, baboons may deduce the existence of higher order structural regularities that link individuals in complex ways, but the gradual emergence of these more complex mental representations can be fully explained as the result of general learning mechanisms and prodigious memories (see Heyes 1994; Wasserman and Astley 1994; Thompson 1995; Schusterman and Kastak 1998 for examples of this view).

In at least one respect the general process argument must be true. It is hard to imagine how a monkey could learn that two other individuals were members of the same martriline except by grouping them together by virtue of their high rates of association. At the same time, general process hypotheses – developed largely in simplified laboratory conditions – seem unable to explain the complexity of behavior observed in nature, where individuals confront an infinitely more complex set of stimuli.

A young baboon, for example, lives in a world where there are thousands of dyadic (and tens of thousands of triadic) relations that must be memorized and organized into groups. The magnitude of the problem makes one wonder whether simple associative mechanisms are equal to the task. Further complicating matters, no single metric defines a "close" social relationship between two individuals, and some relations are transient whereas others persist for much longer times. Moreover, some relationships are transitive (if A is dominant to B and B is dominant to C, then A is always dominant to C) whereas others are not (if female A1 has a close relation with her sister A2 and with male X1, this does not necessarily mean that A2 and X1 have a close bond). Members of the same matrilineal kin group are in some cases mutually substitutable but in other cases retain their distinct identities (recall the example of kin-biased reconciliation described above). Finally, a baboon can belong to many different classes simultaneously, and membership in a class changes often.

In sum, laboratory experiments designed to explain primate social knowledge in terms of a few general-purpose learning mechanisms typically leave out the very complexity they hope to explain. At present we have no proof that general-process mechanisms are wrong, but there are many reasons to believe that these mechanisms are insufficient to explain baboons' social knowledge in all of its complexity (for further discussion, see Cheney and Seyfarth 2007).

An alternative, more innatist view begins with the physiological data reviewed earlier. The existence of specialized areas in the brain that are particularly sensitive to conspecific faces, voices, orientation, movement, gaze direction, and intentions, together with areas in the brain that integrate faces and voices into a unified percept, argues strongly that natural selection has favored innate morphological structures that are functionally specialized to recognize conspecifics, attribute motives to them, and take note of their associations.

Further supporting this view, consider the mechanisms that seem to underlie our experimental results. When a female baboon hears another's female's vocalization, she does not just hear a sound. She perceives a signal that evokes a representation of the caller, what the caller is doing, her rank and family membership. Baboons seem compelled to respond this way. Just as we cannot hear a word without thinking about its meaning, so a baboon cannot hear a vocalization without thinking about the animal who is calling and the events the call describes. And she cannot hear an exchange of vocalizations between, say, Sylvia and Hannah without thinking about these animals' identities, ranks, and family membership, about their relationship, and about its place in the social order. When a baboon hears a sequence of calls that violates the existing social order, she responds within seconds and, as far as we can tell, unconsciously. The speed of her response suggests the existence of a social mind that is innately computational and judgmental.

But while the tendency to make social judgments and form a representation of call meaning may be innate, the *content* of these representations changes all the time. Baboons are always monitoring each other and keeping track of who is consorting with whom, who has fallen in rank, who is moving up, and which families are feuding with each other. Within hours of any societal change, they incorporate this new information into their expectations. They have an innately representational mind that is always open to new information.

In sum, we suggest that, just as humans have an innate predisposition to learn language (e.g., Jackendoff 1994; Pinker 1994) and nutcrackers have an innate predisposition to store and remember the location of seeds (Kamil et al. 1994), so baboons have an innate predisposition to recognize other individuals' ranks and social relationships. We are still a long way, though, from knowing how malleable this predisposition is and what mechanisms underlie it. In this respect, we find ourselves in a position much like that of psycholinguists in the late 1950s, when Chomsky wrote his critique of Skinner's *Verbal Behavior*: we know that the system we are studying is complex and that its development cannot be explained by simple learning mechanisms alone, but we do not yet understand how it develops in the minds of our subjects. To quote Chomsky's (1959) review, without the specific references to language and substituting *baboon* for *human*:

> "As far as acquisition … is concerned, it seems clear that reinforcement, casual observation, and natural inquisitiveness (coupled with a strong tendency to imitate) are important factors, as is the remarkable capacity of the baboon to generalize, hypothesize, and 'process information' in a variety of very special and apparently highly complex ways which we cannot yet describe or begin to understand, and which may be largely innate, or may develop through some sort of learning or through maturation of the nervous system. The manner in which such factors operate and interact … is completely unknown".

7 Do Primates Have Social Theories?

At some point in their lives, all children develop concepts, or theories, about how the world works. Where do these theories come from? Concepts concern features of the world; they provide explanations of why events happen as they do. In this respect, they "embody a systematic set of beliefs that are largely causal in nature" (Keil 1989).

Reviewing studies of children's cognitive development, Keil (1991, 1994) distinguishes between two broadly different hypotheses. "One strongly empiricist account argues that early concepts are devoid of a theory, which then gradually gets overlaid. The other view, which will be called the primal theories account, argues that concepts are embedded within theorylike relations from the start" (1994, p. 235). According to the latter hypothesis, "even in infancy, some very crude theoretical biases start to be abstracted away" from the regularities observed in the social and physical world (p. 238). In Gopnik and Wellman's (1994) terms, an empiricist child may be able to make generalizations that allow her to predict behavior, but these generalizations do not yet constitute a theory because they "are not far removed from the evidence itself." Theories, in contrast, make "predictions about a wide variety of evidence, including evidence that played no role in the theory's initial construction" (1994, p. 261).

Of course, it would go far beyond the evidence to suggest that baboons have the same kind of explicitly articulated theories or concepts we see in young children. Lacking language, it is difficult to imagine how they could or, even if they did, how they would reveal their knowledge to us.

Thus far, we have proposed that baboons have an innately representational mind whose performance, at least in adults, cannot be explained by simple learning mechanisms. We have also suggested that baboons' physiological and cognitive specializations in the social domain have evolved because they need to solve enormously complicated problems very quickly, and we have argued that simple learning mechanisms are not equal to the task. Finally, we have suggested that baboons' classification of each other, however implicit it may be, is based on the application of certain rules of causality as they are applied to other baboons. Do baboons, then, develop during their lives a *theory* of social life?

There are points at which baboons appear to make deductions that go beyond the mere observations of behavior, deductions that suggest that they have some underlying theory about individual motives and the way they are expressed. Consider the ascription of (what we human observers describe as) matrilineal kinship. Here baboons act as if they have a theory: one that is causal because it is based on assumptions about why individuals behave the way they do ("because they are members of the same matriline"), and one that has generality because it makes predictions about a wide variety of evidence, some of which played no role in the theory's initial construction. On the basis of some minimal observations (as yet unknown), baboons conclude that, if A1 and A2 interact in certain ways in some contexts (when feeding, for example), they must be members of the same matriline and can therefore be counted on to behave in predictable ways in many other contexts, i.e., when forming an alliance or grooming, or reconciling with each other's former opponents. In apparently predicting how kin will interact under a wide variety of circumstances, baboons act as if they have a theory about "matrilineal families:" how they should be recognized and how, once identified, individuals within them should behave. The baboons then use this theory to predict behavior.

We suggest, then, that when it comes to recognizing matrilineal kin groups, baboons are "essentialists" (Gelman et al. 1994). They act as if the members of kin groups "have essences or underlying natures that make them the things that they are." This constitutes a kind of theoretical knowledge because "one of the things that theories do is to embody or provide causal linkages from deeper properties to more superficial or surface properties" (Medin 1989).

Consider, as another example, the baboons' ascription of motives to others. When a baboon hears the sequence "Sylvia threat-grunts and Hannah screams," she responds as if she assumes a causal relation between the two events: the threat-grunt and the scream occurred together not by chance but because the former caused the latter. The listener has no other evidence on which to base this assumption because she cannot see the animals interacting. Only her memory of past interactions can guide her perception of current events. And yet she makes the assumption of causality not just for Sylvia and Hannah but for every combination of females whose calls she hears. The listener acts as if she has a general theory about how individuals – *all* individuals - interact.

In much the same way, recall that, in our experiments, a baboon who grooms with X and then, some minutes later, hears X's threat grunt shows little if any response. She acts as if the threat grunt is not directed at her. In contrast, if the baboon has recently fought with X and then hears X's threat grunt, she responds strongly, as if the threat is meant for her. In these situations, once again, the baboon acts as if she has a theory that is applied to every member of her group, not just X. Individuals with whom you groom are kindly motivated toward you and this motivation is likely to persist over time, whereas individuals who have recently threatened you are not kindly motivated and this, too, predicts their subsequent behavior. Of course, these expectations could arise because individuals have formed relatively simple, Pavlovian associations between, say, grooming with X and subsequent friendly behavior with X. But it is also likely, as it is in young children, that as baboons develop into adults some very crude generalizations, or "theoretical biases" (Keil 1994), have begun to be abstracted away from the regularities observed in the social and physical world.

Finally, it is interesting to note that as children's conceptual skills develop they begin to formulate many different concepts in parallel, concepts that can differ widely in both the objects they include and the causal mechanisms that underlie them (Carey 1985). This finding is useful for the psychologists who study children because, as Keil (1989) points out, "no individual concept can be understood without some understanding of how it relates to other concepts." Here again, baboons provide us with a rich source for speculation because they appear to have different social concepts that follow different rules. Individuals grouped together in the same matriline are assumed, by virtue of this grouping, to behave in certain ways toward each other. Individuals arranged in a linear dominance hierarchy follow different rules. The concepts "matrilineal kin group" and "dominance hierarchy" differ in their essential properties, but they can also be merged to form a more complex concept of "ranked matrilines" (Bergman et al. 2003).

8 The Primacy of Social Concepts

Anthropologists and psychologists have often drawn distinctions between two sorts of classification by humans. On the one hand, "biological" classification is typically applied to organisms (plants, animals) other than humans and is based on the observation that, while variation within a species is continuous, variation between species is discrete, and the assignment of an individual to a particular species has important consequences. The classification of one individual as a raccoon and another as a squirrel implies certain immutable properties or "essences" of each that cannot be changed, even if the individual is dyed a different color, born with a birth defect, or altered in some other superficial way. A gray squirrel painted white is still a gray squirrel. Classification of two individuals as members of different species also implies that some activities (like play) can occur between them but others (successful mating) cannot.

Human classifications of other humans are, by contrast, considerably more complex. Some classifications are termed "natural-like" because they depend on inescapable features of an individual that cannot be altered, chosen, or achieved (Hirschfeld 1994). Examples are classification according to age, gender, race, ethnicity, and membership in a particular lineage. Other classifications are termed "social" because they depend on culturally imposed features that can easily be acquired, lost, or changed. Examples are classification according to profession, elected position, and friendship (Hirschfeld 1994).

Several authors have proposed that both the natural and social categories applied by humans to each other have developed, through knowledge transfer, from the biological categories applied by children to plants and nonhuman creatures. As Atran (1990, p. 74) puts it: "Children might initially borrow from their presumptions of the underlying natures of living things in order to better organize their knowledge of humans..." (see also Boyer 1990; Rothbart and Taylor 1990). From a very early age, children recognize that plants and animals vary in a manner that constitutes discrete kinds (e.g., Gelman et al. 1994); they then apply this "essentialist" thinking to their own species. The transfer of categories thus proceeds from the biological to the social. Totemism, in which human groups are associated with animals and differences between animal species symbolize differences between the groups, is often cited as an example of this transfer (Hirschfeld 1994).

Alternatively, Durkheim and Mauss (1903/1963) believe that classification of any sort by humans is a relatively recent phenomenon, and that when early humans first began to organize things into groups they used their own societies as models for such arrangements (see also Needham 1963). As a result, the transfer of categories occurs in the opposite direction, from humans to animals:

> "... the first logical categories were social categories; the first classes of things were classes of men, into which these things were integrated. It was because men were grouped, and thought of themselves in the form of groups, that in their ideas they grouped other things, and in the beginning the two modes of grouping were merged to the point of being indistinct. Moieties were the first genera; clans, the first species."

This view, that the social and psychological concepts we apply to each other are somehow "primary" and that other classifications derive from them, finds some parallel in Carey's (1985) proposal that children begin with a naïve, psychological construal of the world and falsely extend their theories (which work well when applied to humans) to other creatures like dogs or worms (where they often fail). Only later do children realize that biological entities like nonhuman species or plants require a different set of theories from those they apply to humans (see also Keil 1994).

Finally, Hirschfeld (1994) argues against the developmental primacy of either domain. He proposes that there is both a domain-specific competence that underlies humans' essentialist thinking in biology and a domain-specific competence that underlies our essentialist thinking about human social categories. Because the two domains share many similar properties, it is not surprising to find many examples in which thinking in one sphere has been transferred to the other, and vice versa.

At this stage in analysis, our data on baboons have little to say about the extension of knowledge from one domain to another during children's cognitive development. We can, however, adopt a very different perspective and consider the emergence of conceptual thought during human evolution.

Research on baboons raises the possibility that, at some point in history, our ancestors began to recognize and classify members of their own species in ways that were not applied to any other features of their environment. They did so, we suggest, because their socially complex groups created an environment in which an individual's survival and reproduction depended upon her ability to predict other animal's behavior and this, in turn, placed strong selective pressure on individuals to recognize and classify each other (Cheney and Seyfarth 2007). No comparable selective pressure pushed individuals to form comparable classifications of other features of their environment. As a consequence, Durkheim and Mauss were, in evolutionary terms, correct: the first logical categories were social categories, and the first classes of things were classes of conspecifics.

9 Conclusion

Evolution shapes a species' perception and cognition. Among primates, some classes of objects – rocks, clouds – have little effect on the reproductive success of individuals. Subdivisions within these classes are unimportant, so they are perceived and remembered (if they are remembered at all) as generic, undifferentiated groups with no internal structure. Other classes, and the subdivisions within them, have a significant effect on reproductive success. As a result, natural selection has favored individuals whose knowledge of the constituent elements is organized along one or more dimensions. For baboons, the most important features of the environment are other baboons and the crucial dimensions are group membership, dominance rank, matrilineal kinship, and friendships and sexual consortships between males and females.

Compared with those of young children, baboons' theories and concepts are not explicit. Although their social knowledge is implicit, however, this does not make it impoverished.

Baboon social categories constitute representational structures that embody causal explanations of behavior. Baboons recognize other animals' motives and use such attributions to predict their behavior. Their classification of individuals into matrilineal kin groups both links some individuals with others and attributes to these individuals motives and affinities that predict their behavior across a wide variety of contexts. Baboons are intuitive psychologists.

The baboon mind has evolved not just to deal with specific individuals – although it certainly does this – but also to arrange all individuals into social categories that help to predict their behavior. Baboons have a theory of social life because, in large groups where individuals change often but social categories persist, theories and concepts provide the most efficient solution to the problem of predicting behavior.

We do not yet know whether baboon social knowledge is largely innate or develops from experience. Nor can we tell, yet, whether the kind of social categories found in baboons are unique to primates or occur in many other group-living birds and mammals. We can propose, however, that baboon social knowledge sheds light on the evolution of domain-specific social intelligence in humans. Long before they classified other features of the world and developed conceptual knowledge about features outside the social domain, our ancestors classified and reasoned about each other.

References

Atran S (1990) Core domains versus scientific theories: Evidence from systematics and Itza-Maya folk biology. In: Hirschfeld LA, Gelman SA (eds) Mapping the mind. Cambridge University Press, Cambridge, pp. 316–340

Barkow J, Cosmides L, Tooby J (eds) (1992) The adapted mind: evolutionary psychology and the generation of culture. Oxford University Press, New York

Barrett HC, Kurzban R (2006) Modularity in cognition: Framing the debate. Psych Rev 113:628–647

Beehner JC, Bergman TJ, Cheney DL, Seyfarth RM, Whitten PL (2005) The effect of new alpha males on female stress in free-ranging baboons. Anim Behav 69:1211–1221

Bergman TJ, Beehner JC, Cheney DL, Seyfarth RM (2003) Hierarchical classification by rank and kinship in baboons. Science 302:1234–1236

Boyer P (1990) Tradition as truth and communication. Cambridge University Press, New York

Carey S (1985) Conceptual development in childhood. MIT Press, Cambridge

Carey S, Spelke E (1996) Science and core knowledge. Phil Sc 63:515–533

Cheney DL, Seyfarth RM (1997) Reconciliatory grunts by dominant female baboons influence victims' behaviour. Anim Behav 54:409–418

Cheney DL, Seyfarth RM (1999) Recognition of other individuals' social relationships by female baboons. Anim Behav 58:67–75

Cheney DL, Seyfarth RM (2007) Baboon metaphysics: the evolution of a social mind. University of Chicago Press, Chicago

Cheney DL, Seyfarth RM, Silk JB (1995a) The responses of female baboons to anomalous social interactions: Evidence for causal reasoning? J Comp Psychol 109:134–141

Cheney DL, Seyfarth RM, Silk JB (1995b) The role of grunts in reconciling opponents and facilitating interactions among adult female baboons. Anim Behav 50:249–257

Cheney DL, Seyfarth RM, Fischer J, Beehner JC, Bergman TJ, Johnson SE, Kitchen DM, Palombit RA, Rendall D, Silk JB (2004) Factors affecting reproduction and mortality among baboons in the Okavango Delta, Botswana. Intl J Primatol 25:401–428

Chomsky N (1959) A review of B.F. Skinner's *Verbal Behavior*. Language 35:26–58

Cosmides L, Tooby J (1994) Origins of domain specificity: The evolution of functional organization. In: Hirschfeld LA, Gelman SA (eds) Mapping the mind. Cambridge University Press, Cambridge, pp. 85–116

Crockford C, Wittig R, Whiten P, Seyfarth R, Cheney DL (2008) Social stressors and coping mechanisms in wild female baboons. Horm Behav 53:254–265

Dickinson A (1980) Contemporary animal learning theory. Cambridge University Press, Cambridge

Durkheim E, Mauss M (1903/1963) Primitive classification. University of Chicago Press, Chicago

Eifuku S, De Souza WC, Tamura R, Nishijo H, Ono T (2004) Neuronal correlates of face identification in the monkey anterior temporal cortical areas. J Neurophy 91:358–371

Emery NJ, Perrett, DI (2000) How can studies of the monkey brain help us understand "theory of mind" and autism in humans? In: Baron-Cohen S, Tager-Flusberg H, Cohen D (eds) Understanding other minds: perspectives from developmental cognitive neuroscience. 2nd Edn. Oxford University Press, Oxford, pp. 279–310

Engh AE, Hoffmeier RR, Cheney DL, Seyfarth RM (2006a) Who, me? Can baboons infer the target of vocalisations? Anim Behav 71:381–387

Engh AE, Beehner JC, Bergman TJ, Whitten PL, Hoffmeier RR, Seyfarth RM, Cheney DL (2006b) Behavioural and hormonal responses to predation in female chacma baboons (*Papio hamadryas ursinus*). Proc Roy Soc Lond B 273:707–712

Engh AL, Beehner JC, Bergman TJ, Whitten PL, Hoffmeier RR, Seyfarth RM, Cheney DL (2006c) Female hierarchy instability, male immigration, and infanticide increase glucocorticoid levels in female chacma baboons. Anim Behav 71:1227–1237

Fodor JA (1983) The Modularity of mind. MIT Press, Cambridge

Fogassi L, Ferrari PF, Gesierich B, Rozzi S, Chersi F, Rizzolatti G (2005) Parietal lobe: From action organization to intention understanding. Science 308:662–667

Gallistel R, Brown AL, Carey S, Gelman R, Keil F (1991) Lessons from animal learning for the study of cognitive development. In: Carey S, Gelman R (eds) Epigenesis of mind: studies in biology and cognition. Lawrence Erlbaum Associates, Hillsdale NJ, pp. 3–36

Gelman R, Spelke ES, Meck E (1983) What preschoolers know about animate and inanimate objects. In: Rogers D, Sloboda IA (eds) The acquisition of symbolic skills. Plenum, New York, pp. 297–328

Gelman S, Coley JD, Gottfried G (1994) Essentialist beliefs in children: The acquisition of concepts and theories. In: Hirschfeld LA, Gelman SA (eds) Mapping the mind. Cambridge University Press, Cambridge, pp. 341–366

Ghazanfar A, Logothetis N (2003) Facial expressions linked to monkey calls. Nature 423:937–938

Ghazanfar AA, Santos LR (2004) Primate brains in the wild: the sensory bases for social interactions. Nature Rev Neurosci 5:603–616

Gil da Costa R, Braun A, Lopes M, Hauser MD, Carson RE, Herskovitch P, Martin A (2004) Toward an evolutionary perspective on conceptual representation: Species-specific calls activate visual and affective processing systems in the macaque. Proc Natl Acad Sci USA 101:17516–17521

Gopnik A, Wellman H (1994) The theory theory. In: Hirschfeld LA, Gelman SA (eds) Mapping the mind. Cambridge University Press, Cambridge, pp. 257–293

Hauser MD, Andersson K (1994) Left hemisphere dominance for processing vocalizations in adult but not infant rhesus monkeys: Field experiments. Proc Natl Acad Sci USA 91:3946–3948

Hauser MD, Wrangham RW (1990) Recognition of predator and competitor calls in nonhuman primates and birds: A preliminary report. Ethology 86:116–130

Hefner HE, Hefner RS (1984) Temporal lobe lesions and perception of species-specific vocalizations by macaques. Science 226:75–76

Heyes CM (1994) Social cognition in primates. In: Macintosh NJ (ed) Animal earning and ognition. Academic Press, New York, pp. 281–305

Hirschfeld L (1994) Is the acquisition of social categories based on domain-specific competence or on knowledge transfer? In: Hirschfeld LA, Gelman SA (eds) Mapping the mind. Cambridge University Press, Cambridge, pp. 201–233

Hirschfeld L, Gelman S (eds) (1994) Mapping the mind. Cambridge University Press, Cambridge
Hoesch W (1961) Uber ziegen hutende barenpaviane (*Papio ursinus ruacana*). Z Tierpsycho 18:297–301
Humphrey NK (1974) Species and individuals in the perceptual world of monkeys. Perception 3:105–114
Jackendoff R (1994) Patterns in the Mind. Basic Books, New York
Jellema T, Baker CI, Wicker B, Perrett DI (2000) Neural representation for the perception of the intentionality of actions. Brain Cogn 44:280–302
Kamil AC, Balda RP, Olson DJ (1994) Performance of four seed-caching corvid species in the radial-arm maze analog. J Comp Psych 108:385–393
Keil F (1989) Concepts, kinds, and cognitive development. MIT Press, Cambridge
Keil F (1994) The birth and nurturance of concepts by domains: The origins of concepts of living things. In: Hirschfeld LA, Gelman SA (eds) Mapping the mind. Cambridge University Press, Cambridge, pp. 234–254
Kitchen DM, Cheney DL, Seyfarth RM (2005b) Male chacma baboons (*Papio hamadryas ursinus*) discriminate loud call contests between rivals of different relative ranks. Anim Cogn 8:1–6
Kleiner KA (1993) Specific versus non-specific face recognition device. In: de Boysson-Bardies B, de Schonen S, Jusczyk P, MacNeilage P, Morton J (eds) Developmental neurocognition: speech and face processing in the first year of life. Kluwer Academic, Dordrecht, pp 103–108
Leslie AM (1987) Pretense and representation: The origins of "theory of mind". Psych Rev 94:412–426
Medin D (1989) Concepts and conceptual structure. Am Psychol 44:1469–1481
Menzel C (1991) Cognitive aspects of foraging in Japanese monkeys. Anim Behav 41:397–402
Nakahara K, Miyashita T (2005) Understanding intentions: Through the looking glass. Science 308:6444–6445
Needham R (1963) Introduction. In: Durkheim E, Mauss M (eds) Primitive classification. University of Chicago Press, Chicago
Owren MJ, Seyfarth RM, Cheney DL (1997) The acoustic features of vowel-like grunt calls in chacma baboons (*Papio cynocephalus ursinus*): Implications for production processes and functions. J Acoust Soc Am 101:2951–2963
Perrett DI, Harries MH, Mistlin AJ, Hietanen JK (1990) Social signals analyzed at the single cell level: Someone is looking at me, something touched me, something moved. Intl J Comp Psych 4:25–55
Perrett DI, Hietanen JK, Oram MW, Benson PJ (1992) Organization and function of cells responsive to faces in the temporal cortex. PhilTrans R Soc London B 335:23–30
Petersen MR, Beecher MD, Zoloth SR, Moody DB, Stebbins WC (1978) Natural lateralization of species-specific vocalizations by Japanese macaques (*Macaca fuscata*). Science 202:324–327
Pinker S (1994) The language instinct. William Morrow and Sons, New York
Poremba A, Malloy M, Saunders RC, Carson RE, Herskovitch P, Mishkin M (2004) Species-specific calls evoke asymmetric activity in the monkey's temporal poles. Nature 427:448–451
Rendall D (2003) Acoustic correlates of caller identity and affect intensity in the vowel-like grunt vocalizations of baboons. J Acoust Soc Am 113:3390–3402
Rescorla RA (1988) Pavlovian conditioning: it's not what you think it is. Am Psychol 43:151–160
Rizzolatti G, Buccino G (2005) The mirror neuron system and its role in imitation and language. In: Dehaene S, Duhamel JR, Hauser MD, Rizzolatti G (eds) From monkey brain to human brain. MIT Press, Cambridge, pp. 213–234
Rizzolatti G, Craighero L (2004) The mirror-neuron system. Ann Rev Neurosci 27:169–192
Rothbart M, Taylor M (1990) Category labels and social reality: do we view social categories as natural kinds? In: Semin G, Fiedler K (eds) Language and social cognition. Sage, London, pp. 11–36
Schusterman RJ, Kastak DA (1998) Functional equivalence in a California sea lion: relevance to animal social and communicative interactions. Anim Behav 55:1087–1095
Seyfarth, R.M. & Cheney, D.L. 1990. The assessment by vervet monkeys of their own and another species' alarm calls. Anim Behav 40, 754–764

Shettleworth S (1998) Cognition, evolution, and behavior. Oxford University Press, Oxford
Silk JB, Cheney DL, Seyfarth RM (1996) The form and function of post-conflict interactions between female baboons. Anim Behav 52:259–268
Silk JB, Seyfarth RM, Cheney DL (1999) The structure of social relationships among female savannah baboons in Moremi Reserve, Botswana. Behaviour 136:679–703
Silk JB, Alberts SC, Altmann J (2003) Social bonds of female baboons enhance infant survival. Science 302:1231–1234
Silk JB, Altmann J, Alberts SC (2006a) Social relationships among adult female baboons (*Papio cynocephalus*). I. Variation in the strength of social bonds. Behav Ecol Sociobiol 61: 183–195
Silk JB, Altmann J, Alberts SC (2006b) Social relationships among adult female baboons (*Papio cynocephalus*). II: Variation in the quality and stability of social bonds. Behav Ecol Sociobiol 61:197–204
Steiper ME, Young NM, Sukarna TY (2004) Genomic data support the hominoid slowdown and an Early Oligocene estimate for the hominoid–cercopithecoid divergence. Proc Natl Acad Sci USA 101:17021–17026
Teufel C, Hammerschmidt K, Fischer, J (2007) Lack of orienting asymmetries in Barbary macaques: Implications for studies of lateralized auditory processing. Anim Behav 73:249–255
Thompson RKR (1995) Natural and relational concepts in animals. In: Roitblat H, Meyer JA (eds) Comparative approaches to cognitive science. MIT Press, Cambridge, pp. 175–224
Tomasello M (2003) Constructing a language: a usage-based theory of language acquisition. Harvard University Press, Cambridge
Tsao GY, Friewald WA, Knutsen TA, Mandeville JB, Tootell RB (2003) Faces and objects in macaque cerebral cortex. Nature Neurosci 6:989–995
Tsao GY, Friewald WA, Tootell RB, Livingston MS (2006) A cortical region consisting entirely of face-selective cells. Science 311:670–674
Vouloumanos A, Werker, JF (2007) Listening to language at birth: Evidence for a bias for speech in neonates. Dev Sci 10:159–164
Wasserman EA, Astley SL (1994) A behavioral analysis of concepts: Application to pigeons and children. In: Medin DL (ed) Psychology of learning and motivation. vol. 31. Academic Press, New York, pp. 73–132
Weiss DJ, Ghazanfar AA, Miller CT, Hauser MD (2002) Specialized processing of primate facial and vocal expressions: Evidence for cerebral asymmetries. In: Rogers L, Andrews R (eds) Cerebral vertebrate lateralization. Cambridge University Press, New York, pp. 1–74
Wellman H (1990) The child's theory of mind. MIT Press, Cambridge
Wittig RM, Crockford C, Wikberg E, Seyfarth RM, Cheney DL (2007) Kin-mediated reconciliation substitutes for direct reconciliation in baboons. Proc Roy Soc London, Series B 274:1109–1115
Wittig RM, Crockford C, Lehman J, Whiten PL, Seyfarth RM, Cheney DL (2008) Focused grooming networks and stress alleviation in wild female baboons. Horm Behav, doi:10.1016/j.yhbeh.2008.02.009.
Zuberbuhler K (2002b) Causal knowledge in free-ranging Diana monkeys. In: Ghazanfar A (ed), Primate audition, behavior, and neurobiology. CRC Press, Boca Raton, FL, pp. 13–26

What is the Effect of Affect on Bonobo and Chimpanzee Problem Solving?

Brian Hare

Abstract Humans have two close relatives, the bonobo and chimpanzee, whose psychologies differ in ways that will allow for insights into the evolution of our own species. Unfortunately, we know little about bonobos due to their late classification as a species, their scarcity in captivity and the remote location of their natural habitat. Here I review some of the first experimental comparisons of bonobos and chimpanzees, suggesting that their socioecology has shaped their psychology. First, consistent with the observation that bonobos live in more predictable environments, bonobos value future food payoffs less than chimpanzees, while also avoiding the risky foraging decisions that chimpanzees prefer. Second, consistent with the prediction that more predictable environments allow bonobos to be more egalitarian, it was found that bonobos are capable of more flexible cooperative behavior than chimpanzees if the joint problem requires high levels of tolerance. Overall, these comparisons suggest that it is selection on the emotional reactivity of bonobos and chimpanzees that likely played a large role in shaping their differing psychologies, which raises the possibility that changes in human temperament may have also been crucial for the evolution of the unique psychology of our own species as well.

When thinking about human evolution we often encounter comparisons between ourselves and chimpanzees, or *Pan troglodytes*, our closest living relative. However, the species of chimpanzee with which we are all most familiar is only one of our two closest relatives. We actually have two close relatives that are both members of the "chimpanzee" genus, *Pan*, the chimpanzee and the bonobo (*Pan paniscus*). While our own lineage split from the members of the chimpanzee genus *Pan* 5–7 million years ago (Ruvolo 1997), it was only after this split that the two members of the *Pan* genus themselves diverged around 1 million years ago (Won and Hey 2005; see Figure 1A). As a result, both chimpanzees and bonobos share 99.7% of our genome, making both species more closely related to humans than gorillas.

B. Hare
Department of Biological Anthropology and Anatomy, Duke University,
27708 Durham N.C. U.S.A., +1-919-660-7292
e-mail: b.hare@duke.edu

A. Berthoz and Y. Christen (eds.), *Neurobiology of "Umwelt": How Living Beings Perceive the World,* Research and Perspectives in Neurosciences,

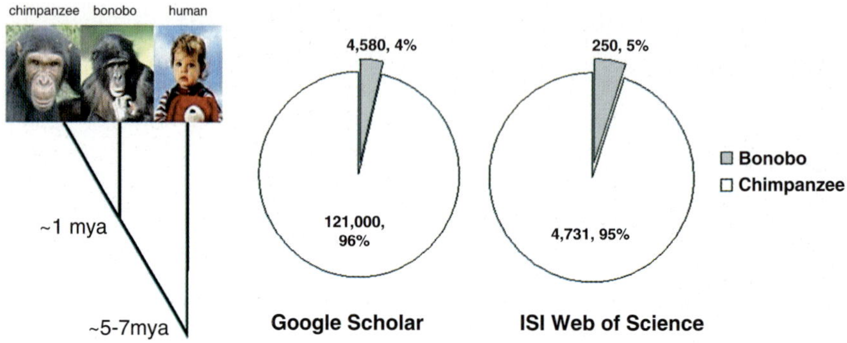

Fig. 1 **a** The phylogeny and estimated divergence times of chimpanzees, bonobos and humans and **b** the absolute number and percentage of citations reported by Google Scholar and ISI Web of science when using either bonobo or chimpanzee as a search term

Our good fortune in having two closest relatives, instead of just one, is often overlooked because our more famous relative, the chimpanzee, has long been the center of attention in science and media alike. This focus is largely due to the fact that bonobos were not even recognized as a species until 1933 and are only found indigenously in one country, the former Zaire or Democratic Republic of Congo. Before the 1970s, hundreds of chimpanzees were taken (or stolen, depending on your perspective) from Africa to fill laboratories, circuses and zoological gardens (the majority from West Africa). Some of these wild-caught animals and their many descendants now live in zoos and labs throughout the developed world. Meanwhile, with little access to the interior of Congo, only a relatively small number of zoos in the U.S, Germany and Belgium have bonobos, approximately two hundred, living in them. Therefore, because of the bonobos' late discovery, remote habitat, and scarcity in captivity, chimpanzees became the relative of choice when comparing the behavior of *Homo* and *Pan*. As a result, instead of being able to split our research efforts more or less equally between our two closest relatives, there has been approximately 20 times more research conducted on chimpanzees than bonobos (see Figure 1B). Due to the paucity of bonobo research, there have been few direct comparisons of bonobos and chimpanzees in any domain of research. This overall skew in the literature led Frans de Waal to dub the bonobo "the forgotten ape" (de Waal and Lanting 1997; Figure 2).

The lack of data on bonobos is particularly problematic when testing phylogenetic hypotheses regarding human evolution. The main goal of comparing humans with chimpanzees is to identify traits that are shared or derived so that we might map out in what ways humans changed since our species split from out last common ancestor with chimpanzees and bonobos. Therefore, when humans possess a trait, say the ability to actively share food, but chimpanzees do not, we feel confident concluding that whatever skill or motivation allows for active sharing in humans must have evolved since the human lineage split (e.g., Jensen et al. 2006). However, while this conclusion may indeed be true, it depends solely on the assumption that our last

a b

Fig. 2 (**a**) Bonobos and (**b**) chimpanzees are our species' two closest living relatives, with both sharing almost 99% of our genome through common descent. This means that the genomes of chimpanzees and bonobos are more similar to humans than that of gorillas. Bonobos and chimpanzees diverged from each other around 1 million years ago and differ in morphology, behavior and perhaps even emotions and cognition in important ways. Bonobos are female dominant, with females forming tight bonds against males through same-sex socio-sexual contact that is thought to limit aggression. In the wild, they have not been seen to cooperatively hunt, use tools or exhibit lethal aggression. Chimpanzees are male dominant, with intense aggression between different groups that can be lethal. Chimpanzees use tools, cooperatively hunt monkeys and will even eat the infants of other chimpanzee groups

common ancestor with chimpanzees and bonobos (or *last common ancestor*) was more chimp-like than bonobo-like. Unfortunately, with no fossil materials available from this ancestor at present, we are left using indirect methods to establish what our common ancestor was like. In agreement with the implicit assumption of many, that the last common ancestor was more chimp-like, some have argued that bonobos are highly divergent and that chimpanzees provide the strongest comparison when identifying derived human traits (e.g., Wrangham and Pilbeam 2001). However, others have argued that, due to certain morphological similarities to humans and lack of climatic fluctuation in the Congo basin, bonobos are likely the species that is more similar to our last common ancestor (de Waal and Lanting 1997). At present there is no definitive proof of which species is more similar to our last common ancestor or whether both species are at the same time highly divergent in some traits but more similar in regard to others (e.g., our last common ancestor possessed a mosaic of bonobo and chimpanzee traits). Therefore, to have the highest level of confidence when identifying derived human traits, it is desirable to examine this trait in both bonobos and chimpanzees. Moreover, because both chimpanzees and bonobos possess certain traits that are more similar to human traits than they are to one another, direct comparisons between bonobos and chimpanzees can provide a powerful method of testing hypotheses concerning the evolutionary process by which traits have changed between species. To demonstrate the potential power of this comparative approach I first contrast the behavioral ecology of bonobos and chimpanzees, which suggests that the two species may differ in their foraging preferences, temperament and cooperative behavior.

1 Bonobo and Chimpanzee Behavioral Ecology

In many ways, bonobos and chimpanzees are highly similar to each other in their behavior and ecology: Both species are predominantly frugivorous and depend on terrestrial herbs when fruit is scarce; they are semi-terrestrial in that they feed and sleep in trees but range across a relatively large area of tropical forest while traveling on the ground; and they live within large communities (potentially >100 members) that often fissure and fuse into various subgroups depending on the availability of sharable food patches found in large fruiting trees. Like humans, but unlike most other primates, the males of both species stay within their natal group whereas it is females that are more likely to immigrate into a neighboring group around puberty ($\sim$9–12 years of age). Finally, the social lives of both species are complicated as they live out life in polygynous multimale-multifemale groups in which mating is promiscuous. Individuals from both species form strong bonds with kin and non-kin within their group, which they must maintain through various affiliative activities so that they might receive support during agonistic interactions (Goodall 1986; Kano 1992; Boesch and Boesch-Achermann 2000).

While it would be easy to concentrate on the many similarities between the two species, there are many striking differences that have also been observed, all of which likely originate from the two species living in geographically distinct habitats for at least one million years (Kano 1992; Boesch et al. 2002). The central difference that has been proposed as driving the speciation of bonobos and chimpanzees is the isolation of bonobos south of the Congo River in forest that allows for a reduction in the intensity of foraging competition due to larger fruit patch size and an increase in the access to terrestrial herbs (White and Wrangham 1988; Wrangham and Peterson 1996; Wrangham 2000). Overall, larger patches of fruit and higher levels of high quality herbs to fall back on when fruit is unavailable reduce the cost of co-feeding and group living for bonobos relative to chimpanzees. Essentially, unlike chimpanzees, bonobos do not need to worry as much about when their next meal will come and whether someone is going to steal it once it does. This difference in food availability across space and time seems to be reflected throughout the social behavior of both species when compared to one another. First, in the wild, it has been observed that reduced foraging competition allows bonobos to form more stable parties (Kuroda 1980; White 1992; Malenky and Wrangham 1994). It has been suggested that these stable parties, in turn, are associated with lower levels of aggression because females more readily form alliances and are able to prevent male aggression to escalate to the level seen in chimpanzees (Wrangham 1986; Wrangham and Peterson 1996; Wilson et al. 2002). Evidence for less aggression in bonobos relative to chimpanzees can be observed in comparing their dominance style and feeding behavior. First, unlike chimpanzees, where each group has an alpha male, within bonobo groups no single male bonobo or coalition of males has ever been observed to dominate a bonobo social group or coerce females with any measure of success (Kano 1992; Parish 1994; Fuirichi 1997; Vervaecke et al, 1999, 2000; Hohmann and Fruth 2003; Paoli et al. 2006). Meanwhile, unlike chimpanzees, where male-male coalitions are ever-prevalent, a comparison across several captive bonobo groups found that female-female coalitions as opposed to either male-male or male-female coalitions are the most prevalent support pattern observed (Stevens et al. 2006). Finally, unlike chimpanzees, where a male can overpower any female, male bonobos who do occasionally attempt to intimidate females - for example, during food sharing episodes - are in some cases attacked by a coalition of females or in many cases simply ignored in favor of other females or kin (Parish 1994; Vervaecke et al. 1999, 2000; Fruth and Hohmann 2002).

Taken together, these findings indicate that, while bonobos and chimpanzees are similar in many ways, there are major differences between the feeding ecologies and social behaviors of these species that seem to be related to one another causally. It seems likely that the high levels of tolerance and lower levels of competition and aggression in bonobos relative to chimpanzees is a result of the richer environment that bonobos inhabit. One prediction of this hypothesis is that the behavioral differences observed in these two species are a reflection of psychological differences that have evolved to allow each species to make adaptive foraging and social decisions based on the constraints found within the environment in which they evolved. If true, how might one test the prediction that the differing socio-ecologies of these two species have shaped their psychologies?

2 Comparing Bonobo and Chimpanzee Psychology

While it seems plausible that the differing feeding ecologies of bonobos may have led to the higher level of tolerance in this species, at first it might seem unclear how to test this causal relationship directly. In fact there is still dispute about exactly how different bonobos and chimpanzee ecology and behavior really are. Some authors emphasize the differences between bonobos and chimpanzees (De Waal and Lanting, 1997; Wrangham and Pilbeam 2001) whereas other authors argue that many of the purported differences (e.g., high levels of tolerance or higher levels of socio-sexual behavior in bonobos) are actually by-products of captivity (Stanford 1998). One potentially powerful way to resolve such debates is with direct quantitative comparisons of chimpanzees and bonobos using experimental methods.

Below I review recent research comparing the foraging behavior of captive chimpanzees and bonobos in two foraging tasks to test whether their feeding preferences in such games match predicted preferences based on how the natural habitats are believed to differ. In addition, I review experiments that test the causal link between these species' temperaments and social behavior by comparing the ability of each to share food and cooperate.

3 Divergent Foraging Psychology in Bonobos and Chimpanzees

Two of the biggest variables that any animal must contend with when finding food are time and risk. First, foraging requires decisions about how long to exploit resources. Animals must frequently decide whether they wish to continue feeding on an immediately available resource or begin searching for another resource that might be more desirable (i.e., less depleted). Second, finding food also requires choices regarding the level of risk animals are willing to accept when foraging. For example, should a primate take a chance of obtaining the most ripe fruit at the branch tips in tree tops where fruit is of highest quality but more dangerous to obtain or should they remain supported by larger limbs and eat lower quality fruit in order to minimize the risk of falling and serious injury?

While all animals must deal with time and risk when foraging, the best foraging strategy will in large part depend on the environmental niche that a species exploits. In some feeding environments, it may in general be advantageous to be more patient and risk averse, whereas in others, a lower return might be received for these same preferences. For example, common marmosets as gumivores likely need more patience when foraging than cotton top tamarins, which are predominantly insectivores. Stevens et al. (2005) tested this prediction with individuals from each species by giving them the choice between two and six pieces of food that were presented simultaneously. Initially, regardless of their choice, they received either option immediately; however, as the test progressed, a delay was introduced that increased across sessions when the larger option was chosen. Both species preferred the larger option if they could have either option immediately, but as delay to obtaining the large food options increased, different preferences developed. Tamarins quickly lost their preference for the larger option when obtaining it required waiting little more than five seconds, whereas marmosets maintained a preference for the larger option even when the delay was twice as long as that which the tamarins were willing to incur. This species difference in patience likely reflects the fact that the psychology of the two species as it relates to foraging evolved under very different conditions. For tamarins, patience does not pay when food is visible (i.e., mobile insect prey), whereas for marmosets it does (immobile gum). Moreover, this research highlights a potentially powerful way to test for differences in the psychology of animals as they relates to suspected differences in their feeding ecology, with the main premise being that, if two species differ in their feeding ecology, then so too should their psychology involved with making foraging decisions (see also Rosati et al. 2006).

Therefore, we used similar techniques to provide an experimental test of the hypothesis that chimpanzees and bonobos evolved in environments with different feeding ecologies. We compared a group of bonobos and chimpanzees in two foraging tasks that were both administered in identical ways to both species. The first task was based on the task used with the callitrichids just described, whereas the second was similar in that it was a choice task but it measured whether subjects preferred to choose a safe or risky payoff. Based on the behavioral ecological data described for bonobos and chimpanzees, we predicted that chimpanzees would be more patient and risk-prone than bonobos in these experiments. Our rationale for these predictions hinged on the fact that bonobos have access to larger feeding patches and high levels of high-quality fall-back food (e.g., herbs) and thus likely evolved in a similar environment where they infrequently gained by taking risks while foraging. Meanwhile, chimpanzees, which must deal with scarcity more frequently and even often participate in the riskiest forms of foraging (i.e., monkey hunting), likely evolved in an environment where they often gained by taking risks while foraging. In support of the ecological prediction, we first found that, when we gave both species the choice of having two pieces of food now or six pieces of food later, chimpanzees maintained their preference for the larger reward with up to a two-minute delay whereas bonobos began to lose their preference after a one-minute delay was introduced (Figure 3). From this differential temporal discounting rate it was concluded that chimpanzees are willing to assign higher value to future foraging payoffs than bonobos (Rosati et al. 2007). In a second task, we compared the two species' risk preferences. Subjects made a series of choices between a fixed option that always yielded four pieces of food when chosen or a variable option that on some trials yielded one food piece but on other trials yielded seven pieces. Although the variable option had two potential payoffs at either extreme, the two food amounts varied with

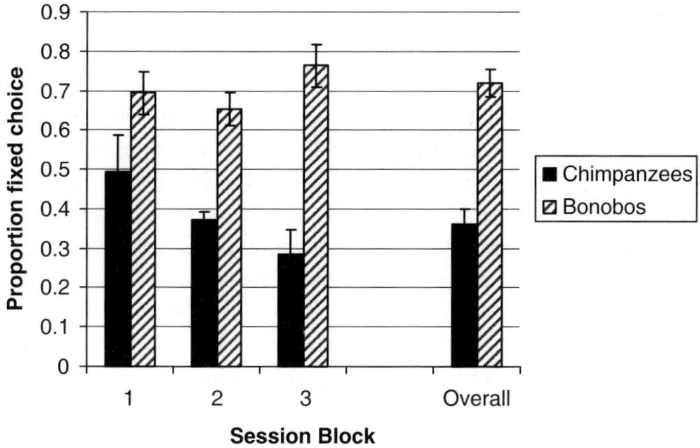

Fig. 3 Patterns of risk preferences in apes across session blocks. Error bars represent standard error. Overall, chimpanzees were significantly risk-seeking, whereas bonobos were significantly risk-averse. Chimpanzees became slightly more risk-seeking as sessions progressed

equal probability. Therefore, the expected values of the two options were equivalent in that a preference for or against either option indicated a sensitivity to risk. Again in support of our predicted species differences, bonobos were risk averse, developing a strong preference for the fixed option, whereas chimpanzees were risk-prone, developing a strong preference for the variable options (Heilbronner et al. 2008).

Taken together, these experimental comparisons between the foraging psychologies of chimpanzees and bonobos suggest that they have very different feeding preferences that map directly onto the differences observed in their feeding ecology. These findings then provide further support to the hypothesis that a variety of differences in bonobo and chimpanzee behavior may have been shaped by the unique challenges their feeding ecologies presented them over evolutionary time. Having established the link between the psychology and feeding ecology of these species, in the next section I discuss a set of test designed to examine whether the differing levels of intra-group competition observed in these two species may have also left a signal in their social psychology.

4 Social Emotions Constrain Primate Cooperation

One of the benefits of living in a social group is that individuals can solve problems with joint action among group mates that any individual alone would be unable to solve. For example, group-living primates often mob predators that are much larger than any individual in the group (e.g., Cheney and Seyfarth 2007). However, while group-living primates frequently act together to defend themselves, their kin or their group mates from predators, non-relatives, or non-group members, the majority of primates in general restrict their cooperative behavior to a few agonistic contexts (Harcourt and de Waal 1992). While naturally occurring cooperation is largely restricted to certain contexts in primates, there is variance among species in their potential to act together in a novel way to solve a novel problem. It has commonly been proposed that there are constraints on cooperative problem solving in primates (Povinelli and O'Neill 2000; Stevens and Hauser 2003; Tomasello et al. 2005), but most recently it has become apparent that species differences in flexible cooperation may in part be a result of constraints created by dominance style. For example, Petit et al. (1992) presented rhesus macaques, with strict dominance hierarchies, and tonkean macaques, with more relaxed hierarchies, with a problem that required joint action to solve it. Sharable amounts of food were placed under heavy rocks that no one monkey could move alone. To obtain the food, at least two monkeys needed to push the stone simultaneously. The more egalitarian tonkean macaques were more successful at producing joint action to retrieve the food than the rhesus, which almost never were able to retrieve the food (only one highly tolerant pair). While the tolerant tonkeans could interact around the stone together, rhesus macaques were simply too inhibited to ever approach the stone while another rhesus was trying to manipulate it. Rhesus macaques could not produce the solution to a novel cooperative problem simply because dominants could not inhibit their aggression towards other group mates that could have potentially helped them solve the problem.

The "emotional reactivity hypothesis" posits that changes in social emotions that accompany changes in social systems and allow species to interact more or less prosocially can dramatically alter social problem-solving skills across species (Hare and Tomasello 2005). The hypothesis led to the prediction that chimpanzees and bonobos with differing emotional reactions to sharing would also differ in how flexibly they could cooperate to obtain food (Hare et al. 2007). Specifically, the hypothesis predicts that, while both species will be skilled at cooperating in novel tasks, the flexibility of chimpanzee cooperation will be more constrained by intolerance between potential partners than in bonobos, since they do not always share food. Indeed, in support of the prediction, it was found that cooperation was constrained in chimpanzee dyads with low social tolerance (Melis et al. 2006a). Removal of such constraints allowed chimpanzees to show relatively sophisticated cooperation. First, highly tolerant pairs who co-fed tended to spontaneously solve the cooperative food-retrieval task. Second, when these same pairs were tested for whether they understood the role of their partner in solving the cooperative task, they spontaneously recruited a conspecific if they needed help in retrieving the food tray. Third, these same pairs preferentially recruited a more skilful partner over a less skilful partner in the same task (Melis et al. 2006b; also see Hirata and Fuwa 2007). Therefore, although chimpanzees can exhibit sophisticated collaborative skills (i.e., flexible recruitment, coordinated and synchronized efforts, etc.), these abilities are not revealed unless tolerance levels between partners are high (Melis et al. 2006a).

To further understand constraints on the evolution of cooperation, Hare et al. (2007) compared the ability of a group of age- and sex-matched bonobos and chimpanzees to cooperatively solve a food-retrieval problem. First, we indexed emotional reactivity by measuring social tolerance while co-feeding in both bonobos and chimpanzees. Dyads were presented with a food platform that had two food dishes spread 2.7 m apart on either end. Food was baited in one of three ways, varying the degree to which it could potentially be monopolized by one individual: two feeding sites with lots of food, one feeding site with lots of food, or one feeding site with two pieces of food. Based on previous observations we predicted that bonobos would co-feed more than chimpanzees (particularly when food was in one dish and easily monopolized) and would actively reduce social tensions while feeding, especially through socio-sexual behavior and play (Kuroda 1980; de Waal 1989; Enomoto 1990; Kano 1992; Furuichi and Ihobe 1994; Parish 1994; Doran et al. 2002; Hohmann and Fruth 2000, but see Stanford 1998). Our comparison revealed that bonobos were more tolerant of co-feeding than chimpanzees. In addition, during co-feeding tests, only bonobos exhibited socio-sexual behavior, and they played more. Thus, regardless of age, chimpanzees showed little socio-sexual behavior or play. Whereas bonobos interacted with ease, chimpanzees appeared to avoid each other.

After confirming experimentally that bonobos are more tolerant of co-feeding as predicted, we conducted a second experiment in which we tested the cooperative ability of bonobos and chimpanzees by presenting them with an instrumental task that required two individuals to simultaneously pull two separate rope ends to obtain out-of-reach food. First, we placed the long food platform out of reach of the testing room, baited it with food, and then threaded a rope through two metal loops at either end of the platform. One end of the rope was then placed within reach of each test room. Pulling only one end of the rope was ineffectual, since the rope would come unthreaded from the loops attached to the platform. Thus, subjects could only obtain the food by pulling both ends of the rope simultaneously towards their room. If two subjects did pull both rope ends, they could pull the tray within reach and obtain the food on the platform. Using this method, we compared the ability of both species to 1) work together to obtain sharable food and 2) work together to obtain a highly monopolizable food reward.

Results support the emotional reactivity hypothesis (Fig. 4). First, the chimpanzees had far more experience solving this same cooperative problem than the bonobos, having participated in previous experiments using this same task (i.e., Melis et al. 2006a,b). In contrast, the bonobos were completely naïve to the task before being tested for this comparison. Yet the bonobos were able to cooperate to obtain highly sharable food at the same level as the chimpanzees and were more skilful than chimpanzees when retrieving highly monopolizable food, regardless of their partner. This was the case when examining differential success both when subjects were initially paired with an opposite sex partner and again when subjects in a second round were repaired with a same-sex partner. The findings support the idea that one route by which social problem-solving can evolve is through selection on emotional systems, such as those controlling the expression of fear and aggression (Hare and Tomasello 2005; Hare et al. 2007). Such flexibility is likely an incidental by-product of selection for social systems and the emotions that allow for them that are adapted for dealing with different levels of feeding competition. Therefore, both bonobos and tonkean macaques more readily show cooperative behaviors than their close, but more despotic relatives.

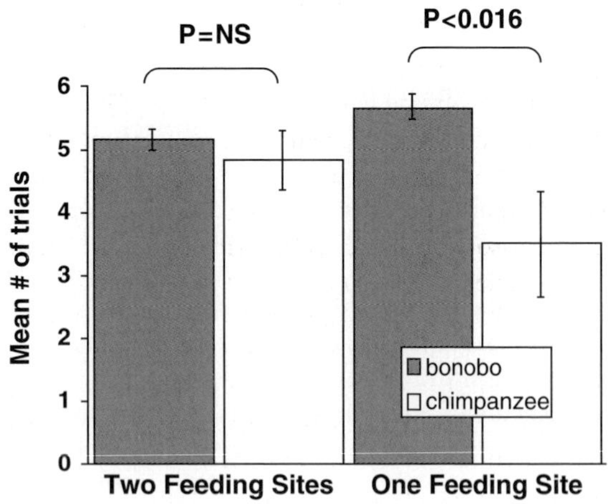

Fig. 4 The mean number (+/− SEM) of trials that pairs of subjects from each species successfully obtained the food platform using cooperation when there were either two feeding dishes or only one

5 Summary and Future Comparisons

Our two closest relatives, the bonobo and chimpanzee, have psychologies that differ in ways that will allow for insights into our own species' evolution. Bonobos live in richer forest and have less intragroup competition over food than chimpanzees. As a result, it has been observed that bonobos in the wild and in captivity are more tolerant in food-sharing contexts than chimpanzees. Based on these differences it has been proposed that 1) each species' psychology will differ, having evolved in response to their differing ecologies and 2) among these psychological differences, it is temperamental differences (levels of emotional reactivity) that are most pronounced and may result in the two species approaching social problems in different ways. First, consistent with the observation that bonobos live in richer environments, bonobos value future food payoffs less than chimpanzees while also avoiding risky foraging decisions that chimpanzees prefer. Second, consistent with lower levels of intergroup competition, it was found that bonobos are capable of more flexible cooperative behavior than chimpanzees if the joint problem requires high levels of tolerance.

Both of these species' differences in foraging preferences and cooperative flexibility support the first proposal above that each species' psychology will differ as a result of their ecology; however, only the finding that tolerance constrains cooperation provides a direct test of the hypothesis that it is temperamental differences that, in part, regulate species differences in problem solving. In this case, we found that cooperation failed in chimpanzees due to social intolerance, even when the two chimpanzees being tested understood that they needed another individual's help to reach their goal. A subordinate might avoid a dominant or a dominant might fail to inhibit her tendency to monopolize a reward. In this way, certain social emotions (elicited during interactions with another animate being) that are normally adaptive in non-cooperative interactions, such as in direct competition over food and mates, potentially limit an individual's or species' behavioural flexibility in approaching novel social problems. Increased behavioral flexibility can result if selection acts on these social emotions so they no longer constrain cooperative interactions. Finally, cognitive evolution can result, if the cognitive ability responsible for the revealed flexibility then itself becomes the target of selection (Hare et al. 2005; Hare 2007).

Therefore, it will be important to consider how the human temperament may be different from other apes and may be in part responsible for human forms of problem-solving (Hare, 2007). However, much more research will be needed if we are to understand the importance of temperamental variables in shaping problem-solving skills. This research will require broad-scale comparisons across a range of species, including non-primates (e.g., Seed et al. 2008; Fidler et al. 2007). In the case of our closest relatives, we will need to compare the emotional reactivity of bonobos and chimpanzees across a range of domains. For example, while we do not have direct evidence yet, future research may reveal that the differing foraging preferences we have found in bonobos and chimpanzees are also regulated by differences in their emotional reactivity. For example, perhaps bonobos in general are more risk averse across a variety of domains due to differences in their emotions' response to uncertainty. An initial comparison of bonobo, chimpanzee and human children in their response to novelty seems to support this possibility (Herrman et al., in preparation).

The future of any such research program involving bonobos and chimpanzees will depend on the large, semi-captive populations of these apes in sanctuaries in Africa. It is only in such sanctuaries that powerful comparisons between bonobos and chimpanzees can be made, because it is only there that there is the necessary sample size. In the two sanctuaries with which we work, there are almost 60 bonobos and 140 -chimpanzees. These sanctuaries provide an unmatched resource not only because of the presence of dozens of young infants of both species that grow up in highly enriched and natural environments (sanctuary apes live together in large social groups in massive tracts of tropical rainforest), but also because they can be tested in a setting similar to a conventional laboratory (indoor enclosures) for a fraction of the cost. By continuing to develop African sanctuaries as world-class resources for researchers, we can look forward to knowing what changed and why during our species' evolution, while simultaneously contributing to the welfare and conservation of the remaining captive and wild populations of our two closest relatives living in their African homes.

Acknowledgments My thanks to Vanessa Woods for providing helpful comments on an earlier version of this manuscript. This research is supported in part by the Humboldt Foundation and the German Federal Ministry for Education and Research.

References

Boesch C, Boesch-Achermann H. (2000) The chimpanzees of the Tai forest. Oxford: Oxford University Press.

Boesch B., Hohmann G, Marchant L. (2002) Behavioral Diversity of Chimpanzees and Bonobos, Cambridge, Cambridge University Press.

Cheney D, Seyfarth R (2007) Baboon metaphysics. Chicago, Chicago University Press.

de Waal F, Lanting F (1997) Bonobo: the forgotten ape. Berkeley, University of California Press.

de Waal FBM, Davis JM (2003) Capuchin cognitive ecology: cooperation based on projected returns. Neuropsychologia 41:221–228.

de Waal F (1989) Behavioral contrasts between bonobo and chimpanzee. In: Marquardt p (ed) Understanding chimpanzees. Cambridge, Harvard University Press, pp. 154–175.

Doran D, Jungers W, Sugiyama Y, Fleagle J, Heesy C (2002) Multivariate and phylogenetic approaches to understanding chimpanzee and bonobo behavioral diversity. In: Boesch C, Hohmann G, Marchant L (eds) Behavioral diversity in chimpanzees and bonobos, Cambridge, Cambridge University Press, pp. 14–34.

Enomoto T (1990) Social play and sexual behavior of the bonobo (*Pan paniscus*) with special reference to flexibility. Primates 31:469–480.

Fidler A, Oers K, Drent P, Kuhn S, Mueller J, Kempanaers B (2007) Drd4 gene polymorphisms are associated with personality variation in a passerine bird. Proc Biols Sci 274:1685–1681.

Fruth B, Hohmann G (2002) How bonobos handle hunts and harvests: why share food? In: Boesch C, Hohmann G, Marchant L (eds) Behavioral diversity in chimpanzees and bonobos, Cambridge, Cambridge University Press, pp. 231–243.

Furuichi T (1997). Agonistic Interactions and Matrifocal Dominance Rank of Wild Bonobos (Pan paniscus) at Wamba. Int J Primatol 18, 855–875.
Furuichi T, Ihobe H (1994) Variation in male relationships in bonobos and chimpanzees. Behaviour 130:211–228.
Goodall J (1986) The chimpanzees of Gombe: patterns of behaviour. Cambridge, MA, Harvard University Press.
Harcourt A, de Waal F (1992) Coalitions and alliances in humans and other animals. Oxford, Oxford University Press.
Hare B (2007) From nonhuman to human mind: what changed and why. Curr Directions Psychol Sci 16:60–64.
Hare B, Tomasello M (2005) Human-like social skills in dogs? Trends Cogn Sci 9:439–444.
Hare B, Plyusnina I, Ignacio N, Schepina O, Stepika A, Wrangham R, Trut L (2005) Social cognitive evolution in captive foxes is a correlated by-product of experimental domestication. Curr Biol 15:226–230.
Hare B, Melis A, Woods V, Hastings S, Wrangham R (2007) Tolerance allows bonobos to outperform chimpanzees in a cooperative task. Curr Biol 17:619–623.
Heilbronner S, Rosati, A, Stevens J, Hare B, Hauser M (2008) A fruit in the hand or two in the bush? Ecological pressures select for divergent risk preferences in chimpanzees and bonobos. Biol Lett 4:246–249.
Hirata S, Fuwa K (2007) Chimpanzees learn to act with other individuals in a cooperative task. Primates 48: 13–21.
Hohmann G, Fruth B (2000) Use and function of genital contact among female bonobos. Anim Behav 60:107–120.
Hohmann G, Fruth B (2003) Culture in bonobos? Between-species and within-species variation in behavior. Curr Anthropol 44:563–609.
Jensen K, Hare B, Call J, Tomasello M (2006) Are chimpanzees spiteful or altruistic when sharing food? Proc Royal Soc B 273:1013–1021.
Kano T (1989) The sexual behavior of pygmy chimpanzees. In: Heltne PG, Marquardt LA (eds) Understanding chimpanzees. Cambridge, Harvard University Press, pp. 176–183.
Kano T (1992) The last ape: pygmy chimpanzee behavior and ecology. Stanford, CA, Stanford University Press.
Kuroda S (1980) Social behavior of the pygmy chimpanzees. Primates 21:181–197.
Malenky R, Wrangham R (1994) A quantitative comparison of terrestrial herbaceous food consumption by *Pan paniscus* in the Lomako Forest, Zaire, and *Pan troglodytes* in Kibale forest, Uganda. Am J Primatol 32:1–12.
Melis A, Hare B, Tomasello M (2006a) Engineering chimpanzee cooperation: social tolerance constrains cooperation. Anim Behav 72:275–286.
Melis A, Hare B, Tomasello M (2006b) Chimpanzees recruit the best collaborators. Science 311:1297–1300.
Paoli T, Palagi E, Borgognini Tarli SM (2006) Reevaluation of dominance hierarchy in bonobos (*Pan paniscus*). Am J Phys Anthropol 130:116–122.
Parish A (1994) Female relationships in bonobos (*Pan paniscus*): evidence for bonding, cooperation, and female dominance in a male-philopatric species. Human Nature 7:61–96.
Petit O, Desportes C. Theirry B (1992) Differential probability of coproduction in two species of macaque (Macaca tonkeana, M. mulatta). Ethology 90:107–120.
Povinelli DJ, O'Neill, DK (2000) Do chimpanzees use their gestures to instruct each other? In: Baron-Cohen S, Tager-Flusberg H, Cohen DJ (eds) Understanding other minds. Perspectives from developmental cognitive neuroscience. Oxford University Press pp. 459–487.
Rosati A, Stevens J, Hauser M (2006) The effect of handling time on temporal discounting in two New World primates. Anim Behav 71:1379–1387.
Rosati A, Stevens J, Hare B, Hauser H (2007) The origins of human patience. Curr Biol 17:1–6.
Ruvolo M (1997) Molecular phylogeny of the hominoids: inferences from multiple independent DNA sequence data sets. Mol Biol Evol 14:248–265.

Stanford CB (1998) The social behavior of chimpanzees and bonobos: empirical evidence and shifting assumptions. Curr Anthropol 39:399–407.

Stanford CB (1999) The social behavior of chimpanzees and bonobos - empirical evidence and shifting assumptions. Curr Anthropol 39:399–420.

Seed A, Clayton N, Emery N (2008) Cooperative problem solving in rooks (Corvus frugilegus). Proc Biol Sci 275:1421–1429.

Stevens J, Hauser M (2004) Why be nice? Psychological constraints on the Evolution of cooperation. Trends Cogn Sci 8, 60.

Stevens J, Hallinan EV, Hauser MD (2005) The ecology and evolution of patience in two new world primates. Biol Lett 1:223–226.

Stevens J, Vervaecke H, de Vries H, van Elsacker L (2006) Social structures in Pan paniscus: testing the female bonding hypothesis. Primates 47:210–217.

Tomasello M, Call J (1997) Primate cognition. Cambridge, Oxford University Press.

Tomasello M, Carpenter M, Call J, Behne T, Moll H (2005) Understanding and sharing intentions: The origins of cultural cognition. Behav Brain Sci 28: 675–691.

Verwaecke H, De Vries H, van Elsacker L (1999) An experimental evaluation of the Consistency of competitive ability and agonistic dominance in different social contexts in captive bonobos. Behaviour 136:423–442.

Verwaecke H, De Vries, van Elsacker L (2000) Function and distribution of coalitions in captive bonobos *(Pan paniscus)*. Primates 41:249–265.

White FJ (1992) Pygmy chimpanzee social organization: variation with party size and between study sites. Am J Primatol 26:203–214.

White FJ, Wrangham RW (1988) Feeding competition and patch size in the chimpanzee species Pan paniscus and Pan troglodytes. Behaviour 105, 148–164.

Wilson ML, Britton NF, Franks NR (2002). Chimpanzees and the mathematics of battle. Proc Royal Soc London 269:1107–1112.

Wrangham RW, Peterson D (1996) Demonic males: apes and the origins of violence. Boston, MA, Houghton Mifflin.

Wrangham RW (1986) Ecology and social relationships in two species of chimpanzee. In Rubenstein DI, Wrangham RW (eds), Ecological Aspects of Social Evolution in Birds and Mammals. Princeton NJ: Princeton University Press, pp 354–378.

Wrangham RW (2000) Why are male chimpanzees more gregarious than mothers? Ascramble competition hypothesis. In: Kappeler P (ed) Primate Males. Cambridge, Cambridge University Press, pp. 248–258.

Wrangham R, Pilbeam D (2001) African apes as time machines. In: All Apes Great and Small. Biruté M.F. Galdikas, Nancy Erickson Briggs, Lori K. Sheeran, Gary L. Shapiro, Jane Goodall (eds) New York, Kluwer Academic/Plenum, pp. 5–18.

Won Y, Hey J (2005) Divergent populations genetics of chimpanzees. J Mol Biol Evol 22:297–307.

Dogs (*Canis familiaris*) are Adapted to Receive Human Communication

Juliane Kaminski

Abstract In recent years, evidence has been accumulating that domestic dogs (*Canis familiaris*) have specialized skills in reading human-given communicative cues (e.g., pointing gestures). These skills seem to be the result of selection pressures during the process of domestication and therefore an adaptation to the dogs' environment, namely human societies. Also, current evidence suggests that dogs' understanding of human gestures is more flexible than was formerly thought. More specifically, dogs distinguish between intended communicative acts and non-intended but target-directed behaviours, suggesting that dogs' behaviour in this domain reflects important aspects of the comprehension of human communicative intentions. However, while children also eavesdrop on communicative interactions between third parties, dogs do not. This can be taken as evidence that dogs take human gestures as directives, while children see them as (sometimes) informative. Also dogs' understanding of gestures seems to be generally more behaviourally based whereas children comprehend gestures in the context of joint attentional interactions.

A species' cognitive adaptations, like its morphological adaptations, reflect the ecological contexts in which it has evolved. Domestic dogs (*Canis familiaris*) have evolved in a very special ecological context. Approximately fifteen thousand years ago, wolves (*Canis lupus*) entered human societies and were domesticated to become one of the most successful species on the planet, the domestic dog (Vila et al. 1997). Since then dogs have been part of human societies and interact with humans in many different ways; they help to hunt, to herd, to protect, etc. (Coppinger and Coppinger 2001). For living in the human world, dogs may have evolved specialized cognitive mechanisms which enable them to interact with human beings and which resemble some of humans' cognitive skills (Hare and Tomasello 2005), making dogs an interesting model for questions regarding the evolution of cognition.

J. Kaminski
Sub-Department of Animal Behaviour, University of Cambridge, High Street, Madingley, Cambridge CB3 8AA UK
e-mail: Kaminski@eva.mpg.de

A. Berthoz and Y. Christen (eds.), *Neurobiology of "Umwelt": How Living Beings Perceive the World,* Research and Perspectives in Neurosciences,

The idea that dogs may indeed have evolved special social skills in interacting with humans comes from research regarding dogs' ability to read humans' communicative gestures. In the so-called object choice paradigm, a human experimenter places a reward (e.g., a piece of food) in one of several identical cups out of view of the subject. After baiting is completed, the human experimenter indicates the location of the reward by producing a communicative gesture (e.g., pointing) towards the correct cup (see for a review Call and Tomasello 2005). Children from early age on understand that, by pointing to one of the locations, the experimenter intends to inform them about where to find the reward (Behne et al. 2005). From an evolutionary perspective, it is striking that our closest living relatives, chimpanzees and other primate species, do not seem to pick up on the same kind of gestures in this kind of communicative interaction. Different primate species do not seem to follow a pointing or gazing gesture, may it be produced by a human (Anderson et al. 1995, 1996; Bräuer et al. 2006; Call et al. 1998) or a conspecific (Call et al. 1998), indicating that there is something about this type of communicative interaction which may be uniquely human (Tomasello et al. 2005).

However, in recent years it has been well established that domestic dogs use all kinds of communicative gestures from humans (see for a review Miklósi and Soproni 2006). Dogs are skillful at using a variety of cues, including pointing, looking and even glancing (Hare et al. 1998; Miklósi et al. 1998; Soproni et al. 2001, 2002). Dogs' performance in this task cannot be explained by olfactory cues because, in a control condition during which the experimenter sat still and did not indicate the location of the food, dogs choose randomly (Bräuer et al. 2006). Also dogs' behavior cannot be explained by learning during the experiment as, in most of the studies, dogs used the gesture from the first trial. Finally, dogs' behaviour cannot be explained by rather low level explanations like, for example, local enhancement as dogs would also follow a human's gestures if the index finger is rather distant from the cup (Soproni et al. 2002), produced statically and even after the human experimenter had actively moved towards and stood still behind one cup while communicating to the other (Hare and Tomasello 1999). The contention that dogs' behaviour in this regard is indeed special and seems to be a result of selection pressures at work during domestication comes from two additional facts.

First, dogs' ancestor, the wolf (*Canis lupus*), does not seem to be as sensitive to human-given gestures as dogs. If tested in comparable settings, wolves fail to use human cues which dogs use readily (Hare et al. 2002; Miklósi et al. 2003). This is not a side effect of being raised in different environments, as even dogs and wolves raised in exactly the same environment differ in their use of gesture and dogs clearly outperform the wolves (Miklósi et al. 2003). Second, dog puppies from a very early age show quite a flexible understanding of human gestures (Hare et al. 2002; Riedel et al., 2008). Six-week-old puppies follow human pointing to a degree which makes low-level explanations (e.g., local enhancement or a simple preference for the human hand) unlikely (Riedel et al.2008). This finding indicates that dogs' ability to use human gestures is not something which dogs need to acquire during their ontogeny. Rather it is an adaptation to life in human societies and, taken together, these findings suggest that selection pressures during domestication have shaped dogs' behaviour in this regard (Hare et al. 2002; Riedel et al. 2008).

However, an important aspect of dogs serving as models for the evolution of human communication is determining which mechanism is involved in dogs' understanding of human forms of communication. Soproni et al. (2001) showed that dogs differentiate between a human gazing at and a human gazing above a target location, indicating that dogs are sensitive to the human's attentional focus during the communicative interaction. However, one important question is to what degree dogs' use of human communication implies a comprehension of human communicative intentions. In interpreting a pointing gesture, humans determine the other individual's intention in directing one's attention towards an outside entity. In addition, when one individual directs signals to another, she makes sure the other knows the communication is for her, with the goal that the other individual knows or does something in response (Sperber and Wilson 1986; Tomasello 2008).

As mentioned earlier, the understanding of others' communicative intention is manifest in humans from an early age. Children from the age of 12 months understand that, by pointing to a certain location, the adult intentionally directed their attention to that bucket (Behne et al. 2005). Children from the same age group clearly distinguished these intended communicative acts from unintended movements which have some of the characteristics of the pointing behaviour. Behne et al. (2005) conducted a condition during which the adult held her hand in a pointing configuration but without ostensive cues. Instead, she pretended to check the time on her watch. Infants did not take this as communicative signals directed to them and choose randomly between both locations (Behne et al. 2005). Children not only use this information when it is communicated to them directly; they also eavesdrop and can extract information from communicative interactions directed at others and not at them (Akhtar 2005; Gräfenhain et al. submitted for publication).

In an attempt to investigate whether dogs, like children, differentiate intended communicative interactions from unintended movements, Kaminski et al. (submitted for publication) replicated Behne et al.'s study and showed that dogs, like children, differentiate between both situations. This is astounding, as it indicates that dogs' understanding of human communication reflects various important aspects of the comprehension of human communicative intentions. The finding that dogs' differentiation between both is not some purely cue-based sensitivity to the eyes comes from a condition during which the human sits with her back turned to the dog. Here the dogs use the communicative gesture provided even though the eyes of the human are not visible at all (Kaminski et al., submitted for publication). Therefore, even though there is much evidence for the important role of the eyes in dog-human interactions (Call et al. 2003; Gacsi et al. 2004; Miklósi et al. 2003; Schwab and Huber 2006; Viranyi et al. 2004), the eyes do not seem to simply function as some kind of trigger which enhances dogs' attention towards the gesture. Instead the eyes seem to be a cue indicating to dogs when communication is directed at them.

However, unlike children, dogs do not follow a gesture if it is directed at another person. In a situation during which the human sits oriented towards them but her gaze alternates between the cup and a person sitting next to her instead of at the dog sitting opposite her, dogs ignore the gesture (Kaminski et al., submitted for publication). This is in contrast to 18-month-old children, who follow a communicative gesture irrespective of the addressee (Gräfenhain et al., submitted for publication). This finding shows that, while children use the information provided by the gesture irrespective of whether it is directed at them, dogs ignore the gesture as long as it is not clearly directed at them (see also Viranyi et al. 2004), indicating that, unlike children, dogs see the gesture more as some kind of directive, telling them where to go. Therefore for them the gesture becomes especially relevant if the addressee is them. From an evolutionary perspective, it is conclusive that dogs have evolved a more selective understanding of human communication. In their everyday interactions with humans, dogs are confronted with all kinds of social cues coming from humans. Among these, the most important are those which are clearly directed at them, hence accompanied by ostensive cues such as eye contact, as these signal that the human actually wants something from them. Dogs' first function was surely to help in activities like hunting and herding, which all require communication over a certain distance (Clutton-Brock 1995; Coppinger and Coppinger 2001). In addition these activities require dogs to obey; therefore a relatively flexible understanding of human directives was surely highly adaptive and, given that dogs are highly dependent on humans (e.g., as a resource for food) was most likely critical for their survival. However, so far we do not know anything about the ontogeny of dogs' eavesdropping. It may be that dogs start out with following gestures irrespective of the addressee. Later, when dogs enter the formal education typical for a domestic dog, this initial tendency may be narrowed down to a more selective use of gestures. Therefore, before any evolutionary conclusions can be drawn, future research is needed.

References

Akhtar N (2005) The robustness of learning through overhearing. Dev Sci 8(2): 199–209.

Anderson JR, Sallaberry P, Barbier H (1995) Use of experimenter-given cues during object-choice tasks by capuchin monkeys. Anim Behav 49: 201–208.

Anderson J, Montant M, Schmitt D (1996) Rhesus monkeys fail to use gaze direction as an experimenter-given cue in an object-choice task. Behav Proc 37: 47–55.

Behne T, Carpenter M, Tomasello M (2005). One-year-olds comprehend the communicative intentions behind gestures in a hiding game. Dev Sci 8: 492–499.

Bräuer J, Kaminski J, Riedel J, Call J, Tomasello M (2006) Making inferences about the location of hidden food: social dog, causal ape. J Comp Psychol 120: 38–47.

Call J, Tomasello M (2005) What chimpanzees know about seeing, revisited: an explanation of the third kind. In: Eilan N, Hoerl C, McCormack T, Roessler J (eds) Joint attention: communication and other minds: issues in philosophy and psychology. New York, Clarendon Press/Oxford University Press, pp. 45–64.

Call J, Hare B, Tomasello M (1998) Chimpanzee gaze following in an object-choice task. Anim Cogn 1: 89–99.

Call J, Agnetta B, Tomasello,M (2000). Cues the chimpanzees do and do not use to find hidden objects. Anim Cogn 3: 23–34.

Call J, Braeuer J, Kaminski J, Tomasello M (2003) Domestic dogs (Canis familiaris) are sensitive to the attentional state of humans. J Comp Psychol 117: 257–263.

Clutton-Brock J (1995). Origins of the dog: domestication and early history. In: Serpell J (ed.) The domestic dog: its evolution, behaviour and interactions with people Cambridge, Cambridge University Press, pp. 7–20.

Coppinger R, Coppinger L (2001) Dogs: a startling nw understanding of anine origin, behavior and evolution. New York, Scribner.

Gacsi M, Miklósi Á, Varga O, Topal J, Csanyi V (2004) Are readers of our face readers of our minds? Dogs (Canis familiaris) show situation-dependent recognition of human's attention. Anim Cogn 7: 144–153.

Hare B, Tomasello M (1999) Domestic dogs (Canis familiaris) use human and conspecific social cues to locate hidden food. J Comp Psychol 113: 173–177.

Hare B, Tomasello M (2005) Human-like social skills in dogs? Trends Cogn Sci 9: 439–444.

Hare B, Call J, Tomasello M (1998) Communication of food location between human and dog (Canis familiaris). Evol Commun 2: 137–159.

Hare B, Brown M, Williamson C, Tomasello M (2002) The domestication of social cognition in dogs. Science 298: 1634–1636.

Miklósi Á, Soproni K (2006) A comparative analysis of animals' understanding of the human pointing gesture. Anim Cogn 9: 81–93.

Miklósi Á, Polgardi R, Topal J, Csanyi V (1998) Use of experimenter-given cues in dogs. Anim Cogn 1: 113–121.

Miklósi Á, Kubinyi E, Jozsef T, Gacsi M, ViranyiZ, Csanyi V (2003) A simple reason for a big difference: wolves do not look back at humans, but dogs do. Curr Biol 13: 763–766.

Riedel J, Schumann K, Kaminski J, Call J, Tomasello M (2008) The early ontogeny of human-dog communication. Anim Behav 75:1003–1014.

Schwab C, Huber L (2006) Obey or not obey? Dogs (Canis familiaris) behave differently in response to attentional states of their owners. J Comp Psychol 120: 169–175.

Soproni K, Miklósi Á, Topal J, Csanyi V (2001) Comprehension of human communicative signs in pet dogs (Canis familiaris). J Comp Psychol 115: 122–126.

Soproni K, Miklósi Á, Topal, Csanyi V (2002) Dogs' (Canis familiaris) responsiveness to human pointing gestures. J Comp Psychol 116: 27–34.

Sperber D, Wilson D (1986) Relevance: communication and cognition. Oxford: Blackwell.

Tomasello M (2008) Origins of human communication. Boston, MIT Press.

Tomasello M, Carpenter M, Call J, Behne T, Moll H (2005) Understanding and sharing intentions: The origins of cultural cognition. Behav Brain Sci 28: 675–735.

Vila C, Savolainen P, Maldonado J, Amorim I. Rice J, Honeycutt R, Crandall KA, Lundeberg J, Wayne RK. (1997) Multiple and ancient origins of the domestic dog. Science 276: 1687–1689.

Viranyi Z, Topal J, Gacsi M, Miklósi Á, Csanyi V (2004) Dogs respond appropriately to cues of humans' attentional focus. Behav Proc 66: 161–172.

What Do Jays Know About Other Minds and Other Times?

Nicola S. Clayton and Nathan J. Emery

Abstract As humans, our thoughts are not "stuck in time." Indeed, it is our ability to mentally dissociate ourselves from the present that allows us to recall the past and plan for our future (mental time travel). We can also reason about what others might be thinking (mental attribution) and thus dissociate ourselves from other selves. Many psychologists have argued that these two forms of mental projection into other times and other minds are unique to humans and that the six-layered prefrontal cortex is the necessary platform for such intelligence. Recent studies challenge these assumptions, however, and some of the most convincing evidence comes not from our closest relatives, the great apes, but from a surprisingly smart, large-brained bird, the western scrub-jay.

Like many other members of the crow family (corvids), these birds hide food for the future and go to great lengths to prevent other birds from stealing their caches. In terms of recalling the past, these birds form complex, highly flexible and integrated memories of the "what, where and when" of specific past caching episodes. They also recall whether another individual was present at the time of caching, and if so, which bird was watching when, and take protective action accordingly, suggesting that they may also be aware of others' knowledge states. This behaviour is only seen in experienced jays that have been thieves themselves in the past, however. Naïve birds that had no thieving experience do not do so, a result that raises the intriguing possibility that experienced jays are able to simulate another bird's viewpoint. Finally, recent work demonstrates that the jays also make provision for a future need and will cache more food in places in which they will not be given breakfast the following morning than in places where they will receive breakfast the

N.S. Clayton
Department of Experimental Psychology, University of Cambridge, Cambridge CB2 3EB, UK
e-mail: nsc22@cam.ac.uk

N.J. Emery
School of Biological & Chemical Sciences, Queen Mary, University of London, London E1 3NS, UK

A. Berthoz and Y. Christen (eds.), *Neurobiology of "Umwelt": How Living Beings Perceive the World,* Research and Perspectives in Neurosciences,

next morning, even though there is plenty of food available to them to consume at the time of making their caching decisions.

We shall argue that, taken together, these results suggest that these birds do possess some knowledge of both other minds and other times, and since birds do not have a six-layered cortical arrangement, these results also challenge the assumption that the mammalian brain is the essential platform for the evolution of these cognitive processes.

1 Introduction

Many psychologists have argued that the ability to travel backwards and forwards in the mind's eye to recollect specific past events and imagine future scenarios is unique to humans. Particularly influential has been the Mental Time Travel hypothesis, in which Suddendorf and Corballis (1997, 2007) have argued that mental time travel is unique to humans and that its emergence was crucial in hominid evolution. According to this hypothesis, animals cannot recall past experiences (episodic memory) or anticipate future states (future planning) because they cannot dissociate themselves from the more or less immediate present; in essence, they are "stuck in time" because they are incapable of temporal perspective taking (Roberts 2002). By this argument, animals learn about the general relationships between stimuli and events from specific episodes without encoding temporal information that enables them to locate these episodes in the past, or when they occurred (e.g., Tulving 1983, 2002; Wheeler 2000).

It has also been argued that humans are unique in being able to reason about what others might be thinking (mental attribution) and thus dissociate their own thought processes from those of others (e.g., Heyes 1998; Saxe 2006; Penn and Povinelli 2007). Indeed, many psychologists have argued that these two forms of mental projection into other times and other minds are two diagnostic features of human cognition. In this chapter, we shall challenge this assumption and argue that food caching by western scrub-jays is an exciting and promising model system for investigating these cognitive processes in animals. Furthermore, we shall argue that much of the success of this model stems from adopting an ethological approach by considering cases in nature in which mental time travel and mental attribution might be important for the animal's survival, paying particular attention to the animal's "Umwelt" or environment, in the broadest sense of the word. For the purposes of our argument, we shall take this to mean how the animal's current ecological conditions and its past evolutionary history (i.e., the role of previous ecological pressures that were imposed upon its ancestors) have shaped the extent to which these two cognitive processes, namely mental time travel and mental attribution, control the animal's behaviour, in our case in terms of the food-caching behaviour of the western scrub-jay.

A major strength of using the food-caching paradigm to test for cognitive abilities lies in the combination of ethological validity and rigorous experimental control, because food-caching is a naturally occurring behaviour, one the jays "do for a living," and consequently they will readily cache and recover food in the laboratory. Indeed an evolutionary analysis of the distribution of caching species among the many members of the corvid family reveals that the ancestor of the corvids was almost certainly a cacher (de Kort and Clayton 2006). Furthermore, unlike many of the standard psychological tests of animal memory, these birds do not need to be trained to cache or recover. Yet, the very fact that memory for food caches can be tested in captive birds allows a level of control that would be difficult, if not impossible, in the field. For example, we can control for the time elapsed between caching and the first opportunity to recover that cache, as well as whether or not the animal can use cues emanating directly from the caches at the time of recovery. In addition, we can test hand-raised birds that have spent their lives in captivity, ones whose reinforcement histories are well documented and whose previous experiences can be experimentally manipulated. This level of control has, in fact, been essential in unravelling questions not only about their ability to remember past caching episodes but also for the experimental tests of whether they can plan for the future, and what they might know about the minds of other jays.

2 Food-Caching by Western Scrub-Jays as a Candidate for Assessing Whether Animals Understand Other Times

A number of animals hide or cache food for future consumption and form accurate, long-lasting memories of where their food caches are located. In laboratory experiments, for example, Western scrub-jays *(Aphelocoma californica)* can accurately remember cache locations for at least 250 days, and Clark's Nutcrackers *(Nucifraga columbiana)* are even more accurate, with no evidence of forgetting for even longer periods of time (Bednekoff et al. 1997). At issue, however, is whether food-caching animals can remember specific past caching episodes in terms of what happened where and, importantly, when (Clayton and Dickinson 1998; Griffiths et al. 1999).

Indeed, we have argued that the ability to remember the what, where and when of unique past episodes is the hallmark of episodic memory that can be tested in animals (Clayton et al. 2001a, 2003a; Salwiczek et al. 2008). Although language-based reports of episodic recall in humans suggest that the retrieved experiences are not only explicitly located in the past but that they are also accompanied by the conscious experience of one's recollections (e.g., Wheeler 2000), of feeling that one is the author of the memory, which is what Tulving (1985) has termed *autonoetic consciousness*, this hypothesis is not something we can test in animals in the absence of agreed non-linguistic markers of consciousness. Clayton and Dickinson (1998) coined the term *episodic-like memory* to refer to the animal's ability to remember what where and when, because we have no way of knowing whether or not this form of remembering is accompanied by the autonoetic consciousness that accompanies human episodic recollections.

2.1 Keeping Track of Perishable Caches

To test whether food-caching jays could remember the what where and when of specific caching events, Clayton and Dickinson (1998) capitalized on one feature of the scrub-jays' ecology, namely, the fact that these birds cache perishable food items, such as worms and other invertebrate prey as well as non-degradable nuts and seeds, and as they do not eat rotten items, recovering perishable food is only valuable as long as the food is still fresh. Consequently, the jays might need to remember not only where they have cached but also which foods are perishable and how long ago they hid them. At this point it is also worth noting that not all food-caching animals hide perishable items. Grey squirrels, for example, avoid the problems of perishability that the jays have by removing the cotyledon from acorns before they cache them, preventing the nuts from germinating and thus spoiling as a food source (Steele et al. 2001).

To test whether these birds remember the what where and when of specific caching episodes, the jays were given a series of trials in which they could cache their preferred food, "wax worms," and the less preferred peanuts in two sand-filled ice cube trays, both of which were made visuo-spatially distinct and trial-unique by attaching Lego Duplo® blocks to the sides of the trays (Clayton and Dickinson 1998). The birds were given the opportunity to cache in different pairs of trays on different trials so that each caching episode was unique. Although the birds had no cue predicting whether or not the wax worms had perished other than the passage of time that had elapsed between the time of caching and the time at which the birds could recover the caches they had hidden previously, the birds rapidly learned that wax worms were fresh when recovered 4 hours after caching, whereas after 124 hours, the worms had decayed. Consequently, the birds avoided the worm caches and instead recovered exclusively peanuts. It was because the animals had been hand-raised, and we therefore knew their precise reinforcement histories, that we could be certain that they had no prior experience of degrading worms.

Having received four pairs of training trials, the birds received a pair of test trials in which the caches were removed prior to recovery and the trays were filled with fresh sand to ensure that the jays could not use any cues emanating directly from the hidden food to choose where to search. The jays searched primarily in the places in which they had hidden the wax worms when the food had been cached 4 hours earlier, but they switched to searching in the peanut sites when the food had been cached 124 hours earlier, suggesting that they did remember what they had hidden in which particular trays and how long ago. Note that the recoveries after both short- and long-retention intervals always occurred at the same time of day (4 hours after caching on the same day as caching or 5 days after caching) and therefore neither circadian rhythms nor the state of hunger at the time of recovery could provide cues to guide the jays' searching behaviour (see de Kort et al. 2005 for further discussion).

Subsequent studies extended these findings by testing whether the jays could also keep track of two perishable food types that decayed at different rates (Clayton et al. 2001b). Again, we appealed to the natural ecology of the birds by reasoning that, in the wild, jays may need to keep track of a number of caches that contain different types of perishable foods, ones that decay at different rates. For example, one might expect worms to rot more rapidly than olives. The jays quickly learned that different foods degrade at different rates and had no difficulty keeping track of which perishable foods had been cached where and how long ago.

We argued that this representation of the time since caching is essential for the efficient recovery of perishable food items, and that Western scrub-jays use a flexible, declarative memory system to do so (Clayton et al. 2001b, 2003b). For example, the jays were also capable of discriminating between episodes in which they had cached the same type of food but in different places and at different times, and thus, on test trials in which no food was actually present at recovery, they searched in those sites in which the worms should still be fresh as opposed to those sites in which the worms would have degraded (Clayton et al. 2001b). Furthermore, when jays received training trials in which the preferred food was found to improve with age rather than deteriorate, the jays switched to searching for the preferred food after the long delay (de Kort et al. 2005), suggesting that the birds were not simply forgetting whether they had cached the preferred food after long time intervals.

Perhaps the most impressive demonstration of flexibility comes from a study in which the jays were allowed to cache perishable and non-perishable items but then discovered, in the interval between caching and recovery, that the perishable food type degraded more quickly than originally thought (Clayton et al. 2003b). This experiment was also informed by considering another ecological factor concerning the "perishability problem" that the jays face, namely that environmental factors influence how quickly a particular item degrades. For example, an item cached in the shade will take longer to rot than one cached in an area that is exposed to direct sunlight, and the caches are more likely to spoil more quickly if there are a spate of sunny days than if there is a cold spell immediately after caching. Indeed, the jays cache perishable foods in an environment where the rate at which foods decay changes across the year, and from day to day, depending on the weather conditions between caching and recovery; so fast in fact, that flexible learning may be essential to their survival. For jays that live in the Central Valley of California (USA), the ambient temperatures rarely fall below 10 °C but may rise to over 40 °C between July and September. At such temperatures, caches that consist of various invertebrates, for example, will degrade rapidly in the heat and more slowly in the cold. So the problem for a scrub-jay is not only to learn how quickly a particular food type degrades but also to be capable of updating information in a flexible manner, based on the ecological conditions that occur in the interim between caching the item and recovering it (de Kort et al. 2005).

Consequently, we reasoned that, if the birds do use a flexible declarative memory system, they should be able to update their knowledge about the rate of perishability of the food and change their search behaviour at recovery accordingly, even though the episodic information about what they cached where and when was encoded prior to the acquisition of the new knowledge about the decay rates.

The jays were able to do just that: if they cached perishable and non-perishable items in different locations in one tray and then subsequently discovered that the perishable items from another tray had degraded more quickly than they expected, then when given the original tray back the birds switched their search preference in favour of the nuts. The birds continued to search for the perishable food if it had been cached recently, thus showing that they had not simply developed a general aversion to searching for food that might perish. To our knowledge, this is the only published demonstration of the declarative flexibility with which animals can update their information after the time of encoding (Clayton et al. 2003b).

3 Food-Caching by Western Scrub-Jays as a Candidate for Assessing Whether Animals Understand Other Minds

Although there are distinct advantages to caching, because food can be hidden for future consumption, this behaviour is both competitive and risky, because other individuals can also steal the caches. The risk of cache theft is significant because up to 30% of caches may be lost each day to pilfering competitors (Vander Wall and Jenkins 2003).

In most food-caching species that have been studied to date, cache theft is opportunistic; either the competitor finds the cache whilst foraging or, if the pilferer happens to witness the caching event, then the onlooker can see where the cacher has hidden the food and displace that individual, so the food can be stolen there and then (see Dally et al. 2006 for a recent review). In the case of scrub-jays and other members of the corvid family, however, cache theft is particularly problematic because, unlike most food-caching animals, these birds have developed an additional thievery tactic, namely, they can rely on their highly accurate observational spatial memory to steal another's caches that they saw being made (Bednekoff and Balda 1996a, b; Heinrich and Pepper 1998; Clayton et al. 2001a). Consequently, the scrub-jays and other corvids can wait until the cacher has left the scene and then steal its caches at will, whenever they are hungry, and without having to rely on successfully displacing a potentially more dominant cacher.

Bugnyar and Kotrschal (2002) suggested that this capacity for observational spatial memory in corvids represented the catalyst for an "evolutionary arms race" between cachers and pilferers, such that pilferers would develop methods for observing cachers as unobtrusively as possible and cachers would come up with strategies to counter the risk of cache pilferage. Furthermore, we have argued that, because these birds act as both the cacher of their own food stores and pilferer of other birds' caches, they have access to both viewpoints, that of the cacher and that of the thief. It is this role-taking that has led to a refinement of increasingly more sophisticated, cognitively based cache protection and pilfering strategies (Emery et al. 2004). Indeed, we have evidence that this refinement can occur very rapidly (i.e., after one experience). When jays were allowed to cache in two trays when watched by an observer, and then they themselves observed one of the trays being pilfered by this same observer (the other tray was in view, but out of reach), they selectively re-cached from the pilfered tray those items that were remaining, and they re-cached these items in places that could not be reached by either a conspecific or experimenter (i.e., in corners of their home cages; Emery et al. 2004). This behaviour suggests that the strategies used by cachers are significantly influenced by the behaviour of potential thieves. As yet, we have not examined the counter-strategies that may be employed by pilferers to increase their chances of stealing caches without reprisal.

Critical to a cacher's use of tactics to counter the risk of cache theft, however, should be the risk that potential thieves pose to their caches, namely whether or not a particular competitor witnessed the caching event and is, therefore, knowledgeable as to cache locations (Clayton et al. 2007). In short, what we are arguing is that the very nature of the cache-pilfering and cache-protection system in these birds has led to the development of complex cognitive processes that may be akin to mental attribution (Emery and Clayton 2004b, 2008). What evidence do we have for making such a controversial claim?

3.1 Keeping an Eye on The Competition: The Cache Protection Strategies of Scrub-Jays

In the first set of experiments, we tested whether the jays were sensitive to what a competitor can and cannot see, using a competitive foraging paradigm similar to that which Hare and colleagues (2000, 2001) developed for testing mental attribution by chimpanzees (Dally et al. 2005; Fig. 1). One bird, the cacher, was given the opportunity to hide food in two different, visuo-spatially distinct trays that were placed in its home cage. One of these trays could easily seen by another jay, the observer in the next cage, but the other tray was placed behind a barrier such that it was out of view of the observer. Both trays were equidistant from the observer's cage. Not surprisingly, the cachers showed a clear preference to hide their food in the tray

Fig. 1 The barrier experiment (Dally et al. 2005). Scrub-jays are presented with two trays; one hidden behind a barrier out of view of an observer and a second in sight of an observer. Cachers predominantly place their caches in the "out of sight" tray, but only when they are observed caching. They cache equally in both trays when they cache in private and there is no potential for cache theft. Drawing by Scott Stevens

that was placed behind a barrier when the observer was watching them. However, when the observer's view of the entire caching event was obscured by a screen so that all caching events were out of the observer's view, the cacher showed no such preference, hiding approximately equal numbers of food items in both trays.

We reasoned that, if the jays can understand what competitors can and cannot see, they should be able to capitalize on relative differences in the level of visual access that observers have to cache sites. By caching in sites that observers might find hard to see, such as those in the shade, cachers might reduce the quality and transfer of visual information to observers, thereby making the location of cache sites less certain. To determine whether scrub-jays would exploit hard-to-see sites, we examined the effects of sun and shade on a scrub-jay's choice of where to cache (Dally et al. 2004; Fig. 2). As in the previous experiment, the jays were given the opportunity to cache in two trays that were equidistant from the observer's cage, but the barrier was removed and a lamp was placed close to each of them. By switching one of the lamps on, and leaving the other switched off, we were able to create a sunny site that was four times as bright as the shady site. When the observer jay could witness the caching event, our cachers were selective, hiding most of their food in the shady tray, but when caching was made in private they placed approximately equal numbers of food items in both caching trays.

Subsequent experiments established that the jays also use distance as a cache-protection strategy, preferring to hide food in a tray furthest from the observer when that bird could witness the caching event, yet caching equally in both the near and the far tray when the observer's view was obscured so that they could cache in private (Dally et al. 2005). Taken together, these results suggest that, like chimpanzees,

Fig. 2 The sun/shade experiment (Dally et al. 2004). Scrub-jays are presented with two trays; one in a darkened part of the cage, a second in a brightly lit part of the cage. Cachers predominantly place their caches in the tray located in the darkened part of the cage, but only when they are observed caching. They cache equally in both trays when they cache in private and there is no potential for cache theft. Drawing by Scott Stevens

scrub-jays are sensitive to what a competitor can and cannot see, exploiting a range of environmental variables, including the level of ambient light and the relative distance of cache sites to the observer, to reduce the transfer of visual information available to the potential pilferer.

3.2 Keeping Track of Who was Watching When

In the second set of experiments, we tested whether the jays could keep track of which particular individual was watching them caching during specific caching episodes (Dally et al. 2006a). These experiments were also shaped by thinking about the jays' natural ecology, in this case the fact that the potential pilferers do not necessarily leave the area in which the cacher has cached, or that whilst the same observers are not always present, there is always another bird in the immediate vicinity. The cachers were given the opportunity to hide food in one of two visuo-spatially unique caching trays in the presence of two different observers in two consecutive caching events (Observer A watches caches being made in Tray A, and Observer B watches caches being made in Tray B), and then to recover their caches either in private, when watched by one of the two observers, or when observed by a "control" bird that had not witnessed either caching event.

When the jays were given the opportunity to recover their caches in private, they re-cached about 50% of the items they had hidden, moving them to new sites. In the presence of a prior observer, the cachers also re-cached the food items, but their pattern of re-caching was very different from that we saw when they re-cached in private. In the presence of an observer, they repeatedly moved the caches that particular observer had seen them make, re-hiding them in the original caching tray. By moving caches repeatedly, cachers may reduce the accuracy with which the observer would be able to steal their caches, rather like the children's shell game. Similarly very few caches were moved during the "control" condition. Presumably this is because moving these items would only have provided observers, who were currently ignorant of their existence, with information that they might have been used to facilitate future cache theft. We have argued that, taken together, these findings suggest that the jays do keep track of which particular individual was watching during particular caching events (but see Clayton et al. 2007 for further discussion).

These findings raise questions about whether the scrub-jay has a "theory of mind," given that the jays appear to be sensitive to what particular individuals have and have not seen when deciding which caches to protect. Of course, as we pointed out in the original paper (Dally et al. 2006b), we caution that these abilities do not necessarily require a human-like theory of mind. For one thing, it is hard to imagine how a non-linguistic subject could theorize about the minds of other individuals. Clearly, it would be informative to develop a model of how the jays might achieve this seemingly complex behaviour without language (see Emery and Clayton 2008 for a recent attempt to construct a cognitive architecture of mind-reading by scrub-jays). However, at the very least, the selectivity of the cache-protection behaviours does appear to depend on sensitivity to what others have and have not seen. In short, these studies show that scrub-jays keep an eye on the competition and protect their caches accordingly. Such behaviour would appear to meet the behavioural criteria for one form of theory of mind, namely knowledge attribution, if by the term we mean the ability to attribute different informational states to particular individuals.

3.3 Experience – Projection: It Takes a Thief to Know One

There is one particularly striking finding about the re-caching behaviour of these birds and that is that not all Western scrub-jays engage in it. Emery and Clayton (2001) found that re-caching behaviour depends not only on whether or not the cacher was observed by another jay during caching but also upon the previous experience of being a pilferer. Whereas experienced thieves engaged in high levels of re-caching at recovery when they were observed during the previous caching episode, control birds, who had not been thieves in the past and therefore had no prior experience of stealing other birds' caches, showed hardly any re-caching at all.

The fact that only experienced jays re-cache has a number of important implications. The first is that this behaviour cannot be innate, otherwise all the birds should re-cache. Importantly, we can also rule out a simple conditioning explanation because the birds never received any positive reinforcement or any punishment for re-caching, given that they never had the opportunity to learn about the fate of the caches that they had re-cached. As a consequence, we can make the inference that the jay used information gained during the previous caching event to anticipate whether or not its caches were likely to be stolen and thus engaged in the appropriate cache-protection strategy at recovery, namely whether or not to re-cache.

Emery and Clayton (2004b) have argued that the fact that experienced birds differ so dramatically from control birds who lack the experience of being a thief suggests that the experienced jays are not only capable of protective action against future theft but also experience projection. Experience projection refers to a second form of theory of mind, namely the ability to use one's own experiences – in this case of having been a thief - to predict how another individual might think or behave – in this case what the potential pilferer might do. Experience projection has yet to be demonstrated in any of the great apes, other than humans. Consequently, most people have assumed that experience projection was a uniquely human trait. The jay studies challenge this assumption. The fact that experienced jays re-cache has one further implication: the jays may be sensitive to future needs. After all, the only benefit of re-caching the food at recovery is to increase the likelihood that they will then be able to consume those caches at a later date. In fact, we now have good evidence that these jays can plan for the future (Correia et al. 2007; Raby et al. 2007).

Fig. 3 The planning for breakfast experiment (Raby et al. 2007). Scrub-jays are first trained that one of the rooms contains non-cacheable food in the morning (breakfast room) whereas a second room does not contain any food in the morning (non-breakfast room); food is then freely available for the rest of the day for six days. The birds are then presented with a novel test on the evening of the sixth day, namely a bowl of whole food items that they can cache in either room. The scrub-jays cache predominantly in the non-breakfast Room, the room in which they do not expect to find food the next morning. Drawing by Scott Stevens

3.4 Planning for Tomorrow's Breakfast

We tested whether the jays could plan for a future motivational need as opposed to a current one, namely for tomorrow's breakfast (Raby et al. 2007; Fig. 3). To do so, each jay was housed in a three-room compartment in which they could eat but not cache food, because the food was always provided in a powdered form. Each morning, for six days, the jays woke up in one of the two end compartments, such that they received equal amounts of experience with both of them. In one compartment, the jays were always given breakfast and in the other they were not. After this training, the jays were unexpectedly given food to eat and cache in the evening. We reasoned that, if they were capable of thinking about the future, they should cache more food in the compartment in which they had not been given breakfast and therefore would expect to be hungry the next morning, relative to the compartment in which they had been given breakfast. This they did, spontaneously and without prior training.

4 Are Western Scrub-Jays Unique?

We recognise that, by claiming that Western scrub-jays may have episodic memory, future planning and mental attribution, we will be viewed with scepticism by many comparative and human psychologists, especially those who adhere to the ideas that human cognition is unique or that animals do not have minds. Although we have been wary of this throughout our experimental program, and have attempted to control for alternative explanations for our results, we do not think that agreement on these issues will be forthcoming any time in the near future. However, we believe that, aside from the results, there are good biological reasons for why Western scrub-jays and possibly other animals may have these abilities. However, at first glance, our results present a paradox. How can such abilities be present in a bird that is distantly related to us and in one whose brain is the size of a walnut?

First, evolutionary relatedness to humans does not necessarily imply similar cognitive mechanisms. This out-dated scala naturae view on the evolution of cognition does not follow Darwinian principles (Hodos and Campbell 1969). Evolution does not progress through time like a ladder, with "lower" animals such as insects, fish and reptiles on the bottom rungs and "higher" animals such as birds, mammals and ultimately humans on the top rungs. Neither do animals from one group simply evolve into new animals. Rather speciation spreads like the branches of a tree, in which species adapt to new environments and face new challenges and so change as a result. All species therefore have a common ancestor. The questions in relation to the evolution of cognition therefore become: at what point in the evolutionary tree did the ability arise (common ancestor) or, if the ability is shared by many species, has the ability arisen through common ascent (homology) or convergence (analogy)? If the trait is homologous, then we expect that most of the species from the common ancestor upwards should share it. For example, as all animals will habituate to repeated presentations of the same stimuli, the appearance of habituation learning in distantly related species must be homologous. By contrast, at the present time, we only have evidence that Western scrub-jays and humans (and possibly apes) can plan for the future. As their last common ancestor lived around 300 millions years ago, cognition in corvids and apes have most likely arisen though convergent evolution. In the case of apes and humans, this may also be a case of homology. We have argued that similar cognitive abilities in apes and corvids are a clear example of evolutionary convergence (Emery and Clayton 2004b), which may be the result of similar selection pressures from the complexity of the social and physical environments of these two families, which may have also influenced other animal groups, such as elephants, killer whales and dolphins, which have similar abilities (Emery 2006; Fig. 4).

If common descent is not necessary for the appearance of complex cognition, as we suggest, then a second problem arises as to how these jays with their tiny brains can possibly possess supposedly human unique cognitive abilities? We already know that, although avian brains are smaller in absolute size, there are major size differences within *Aves*. Corvids and parrots, for example, have brains which are relatively the same size as the great apes (Emery and Clayton 2004a) and, interestingly, the Western scrub-jay brain is relatively the same size as the *Australopithecus* brain (Pravosudov and de Kort 2006). However, until recently, the question of neural complexity would have seemed an insurmountable problem, as the avian brain was suggested to be largely comprised of striatal areas, such as the neostriatum and hyperstriatum, structures in the basal ganglia, and so their function restricted to instinctive, species-typical behaviour, such as feeding, parenting and mating. We

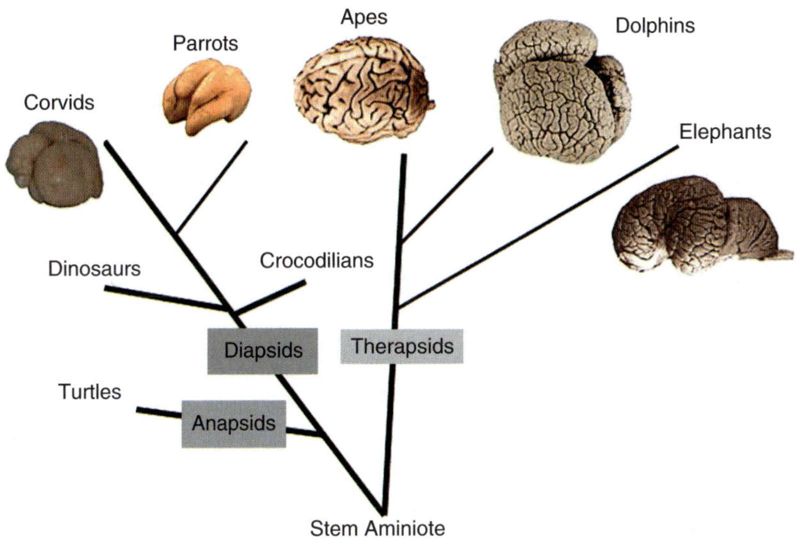

Fig. 4 Corvids and apes are examples of convergent evolution of cognition, but with some degree of divergence in neuroanatomy (Emery and Clayton 2004a). As the last common ancestor (stem amniote) of birds and mammals lived around 300 million years ago, any similarity in their cognitive abilities must have arisen through convergence, especially as most species within the mammalian and avian lines do not display similar abilities, and there is no evidence that other ancestral groups also possess or possessed these skills. Other potential groups to converge on the same abilities are parrots, cetaceans and elephants

now know through studies of anatomical connectivity, neurochemistry and function in avian brains that these areas were wrongly named, as they are more cortical-like in structure and function. A recent nomenclature change to the entire avian brain has redressed this imbalance by renaming these supposed striatal areas as derived from the pallium, in the same way as the mammalian neocortex (Jarvis and Consortium 2005). Therefore, the nidopallium (neostriatum) and mesopallium (hyperstriatum) are analogous to the neocortex of mammals (Emery and Clayton 2005), and even the caudalateral nidopallium has been suggested to be equivalent to the mammalian prefrontal cortex (Güntürkün 2005). Receptors for neurotransmitters, such as dopamine, and neurochemical pathways involved in cognitive processing in mammals are also similarly distributed in the avian brain. In addition, there have even been suggestions that avian and mammalian brains are similar in the connectivity patterns of sensory (vision, touch) and motor neuronal circuits (Jarvis and Consortium 2005), as well as circuits involved in vocal learning (Jarvis et al. 2000). This hypothesis inevitably leads to the further possibility that circuits involved in cognitive processing are also similar. So even though bird brains are much smaller than typical mammalian brains, and tens of times smaller than the ape or human brain, there is good evidence that absolute size does not matter (similar relative brain size); it's what's done with it that counts (similar circuitry and neurochemistry).

References

Bednekoff PA, Balda RP (1996a) Observational spatial memory in Clark's nutcrackers and Mexican jays. Animal Behav 52: 833–839.

Bednekoff PA, Balda RP (1996b) Social caching and observational spatial memory in pinyon jays. Behaviour 133: 807–826.

Bednekoff PA, Balda RP, Kamil AC (1997) Long-term spatial memory in four seed-caching corvid species. Animal Behav 53: 335–341.

Bugnyar T, Kotrschal K (2002) Observational learning and the raiding of food caches in ravens, Corvus corax: is it 'tactical' deception? Animal Behav 64: 185–195.

Clayton NS, Dickinson A (1998) Episodic-like memory during cache recovery by scrub jays. Nature 395: 272–274.

Clayton NS, Griffiths DP, Emery NJ, Dickinson A (2001a) Elements of episodic-like memory in animals. Phil Trans Royal Soc London B 356: 1483–1491.

Clayton NS, Yu KS, Dickinson A (2001b) Scrub-jays (*Aphelocoma coerulescens*) form integrated memory for multiple features of caching episodes. J Exp Psychol: Animal Behav Processes 27: 17–29.

Clayton NS, Bussey TJ, Dickinson A (2003a) Can animals recall the past and plan for the future? Nature Rev Neurosci 4: 685–691.

Clayton NS, Yu KS, Dickinson A (2003b) Interacting cache memories: evidence for flexible memory use by scrub jays.J ExpPsychol: Animal Behav Processes 29: 14–22.

Clayton NS, Dally JM, Emery NJ (2007) Social cognition by food-caching corvids: The western scrub-jay as a natural psychologist. Phil Trans Royal Soc London B 362: 507–522.

Correia SPC, Dickinson A, Clayton NS (2007) Western scrub-jays anticipate future needs independently of their current motivational state. Curr Biol 17: 856–861.

Dally JM, Emery NJ, Clayton NS (2004) Cache protection strategies in western scrub-jays (*Aphelocoma californica*): hiding food in the shade.Proc Royal Soc London B: Biol Lett 271: S387–S390.

Dally JM, Emery NJ, Clayton NS (2005) Cache protection strategies in western scrub-jays: implications for social cognition. Animal Behav 70: 1251–1263.

Dally JM, Emery NJ, Clayton NS (2006a) Food-caching scrub-jays keep track of who was watching when. Science 312: 1662–1665.

Dally JM. Clayton NS, Emery NJ (2006b) The behaviour and evolution of cache protection and pilferage. Animal Behav 72: 13–23.

de Kort SR, Clayton NS (2006) An evolutionary perspective on caching by corvids." Proc Royal Soc London B 273: 417–423.

de Kort SR, Dickinson A, Clayton NS (2005) Retrospective cognition by food-caching western scrub-jays. Learn Motivation 36: 159–176.

Emery NJ (2006) Cognitive ornithology: the evolution of avian intelligence. Phil Trans Royal Soc London B 361: 23–43.

Emery NJ, Clayton NS (2001) Effects of experience and social context on prospective caching strategies by scrub jays. Nature 414: 443–446.

Emery NJ, Clayton NS (2004a) Comparing the complex cognition of birds and primates. In; Rogers LJ, Kaplan g (eds) Comparative vertebrate cognition: are primates superior to non-primates? New York, Kluwer Academic Press, pp. 3–55.

Emery NJ, Clayton NS (2004b) The mentality of crows: Convergent evolution of intelligence in corvids and apes. Science 306: 1903–1907.

Emery NJ, Clayton NS (2005) Evolution of avian brain and intelligence. Curr Biol 15: R946–R950.

Emery NJ, Clayton NS (2008) How to build a scrub-jay that reads minds. In: Itakura S, Fujita K (eds) Origins of the social mind: evolutionary and developmental views. Kyoto, Japan, Springer Japan, in press.

Emery NJ, Dally JM, Clayton NS (2004) Western scrub-jays (*Aphelocoma californica*) use cognitive strategies to protect their caches from thieving conspecifics. Animal Cogn 7: 37–43.

Griffiths DP, Dickinson A, Clayton NS (1999) Episodic memory: what can animals remember about their past? Trends Cogn Sci 3: 74–80.

Güntürkün O (2005) The avian 'prefrontal cortex' and cognition. Curr Opin Neurobiol 15: 686–693.

Hare B, Call J, Tomasello M (2000) Chimpanzees know what conspecifics do and do not see. Animal Behav 59: 771–785.

Hare BJ, Call J, Tomasello M (2001) Do chimpanzees know what conspecifics know? Animal Behav 61: 139–151.

Heinrich B, Pepper JW (1998) Influence of competitors on caching behaviour in the common raven, *Corvus corax*. Animal Behav 56: 1083–1090.

Heyes CM (1998) Theory of mind in nonhuman primates. Behav Brain Sci 21: 101–148.

Hodos W, Campbell CBG (1969) Why there is no theory in comparative psychology? Psychol Rev 76: 337–350.

Jarvis ED, Consortium ABN (2005) Avian brains and a new understanding of vertebrate brain evolution. Nature Rev Neurosci 6: 151–159.

Jarvis ED, Ribeiro S, Vielliard J, DaSilva M, Vebntura D, Mello CV (2000) Behaviorally driven gene expression reveals song nuclei in hummingbird brain. Nature 406: 628–632.

Penn DC, Povinelli DJ (2007) On the lack of evidence that non-human animals possess anything remotely resembling a 'theory of mind'. Phil Trans Royal Soc London B 362: 731–744.

Pravosudov VV, de Kort SR (2006) Is the western scrub-jay (Aphelocoma californica) really an underdog among food-caching corvids when it comes to hippocampal volume and caching propensity? Brain, Behav Evol 67: 1–9.

Raby CR, Alexis DM, Dickinson A, Clayton NS (2007) Planning for the future by western scrub-jays. Nature 445: 919–921.

Roberts WA (2002) Are animals stuck in time? Psychol Bull 128: 473–489.

Salwiczek LH, Dickinson A, Clayton NS (2008) What do animals remember about their past? In: Cognitive Psychology of Memory. In R. Menzel (Ed.), Learning Theory and Behavior. Vol. [1] of Learning and Memory: A Comprehensive Reference, 4 vols. (J. Byrne Ed.), pp. 441–460. Elsevier: Oxford.

Saxe R (2006) Uniquely human social cognition. Curr Opin Neurobiol 16: 235–239.

Steele MA, Turner G, Smallwood PD, Wolff JO, Radillo J (2001) Cache management by small mammals: experimental evidence for the significance of acorn embryo excision. J Mammol 82: 35–42.

Suddendorf T, Corballis MC (1997) Mental time travel and the evolution of the human mind. Genet Soc Gen Psychol Monog 123: 133–167.

Suddendorf T, Corballis MC (2007) The evolution of foresight: What is mental time travel, and is it unique to humans? Behav Brain Sci 30: 299–351.

Tulving E (1983) Elements of episodic memory. Oxford, UK, Clarendon Press.

Tulving E (1985) Memory and consciousness. Can Psychol 26: 1–12.

Tulving E (2002) Chronesthesia: awareness of subjective time. In: Stuss DT, Knight RC (eds) Principles of frontal lobe functions. New York, Oxford University Press, pp. 311–325.

Vander Wall SB, Jenkins SH (2003) Reciprocal pilferage and the evolution of food-hoarding behavior. Behav Ecol 14: 656–667.

Wheeler MA (2000) Episodic memory and autonoetic awareness. In: Tulving E, Craik FIM (eds) The Oxford handbook of memory. New York, Oxford University Press, pp. 597–608.

Blind as a Bat? The Sensory Basis of Orientation and Navigation at Night

Richard Holland

Abstract Animals that move around after dusk face many challenges but none as great as the need to be able to orient and navigate in low light levels. Bats are most famous for their adaptation to the nocturnal niche and have developed one of the most sophisticated biological sonar systems known. This echolocation allows them to capture small insects on the wing, sometimes in dense forests with high accuracy. Indeed in tests it is impossible to distinguish between a bat that uses echolocation to detect objects and one that uses vision. Despite the fact that echolocating bats produce some of the loudest sounds known to man, echolocation is only of very short range however. Being wide ranging animals capable of flight, bats often move beyond the useful limits of echolocation and must call on a variety of sensory cues to be able to find there way round in the dark, to return to a home roost or to migrate from breeding to wintering grounds. It turns out that "blind as a bat" is a misnomer in every sense of the word. Not only is vision important to these animals but a number of other senses are called upon to allow them to return home after a night of foraging for insects. It is not just bats that face this challenge. Many songbirds migrate at night to allow them to use more favourable atmospheric conditions and to avoid predators. In this chapter I will discuss the theory and the evidence of how animals such as bats and songbirds perceive the world around them at night for the purpose of finding their way home, whether it be a journey of 20 km to return to a home roost after a night of hunting insects or a migratory journey of 1000s of km across continents. This task is achieved using sensory systems in ways that humans can only imagine, and nowhere is the term "Umwelt" more appropriate.

R. Holland
Marie Curie Outgoing International Fellow, Institute for Integrative and Comparative Biology, University of Leeds, Leeds, LS2 9JT, UK
e-mail: bgyraho@leeds.ac.uk

A. Berthoz and Y. Christen (eds.), *Neurobiology of "Umwelt": How Living Beings Perceive the World,* Research and Perspectives in Neurosciences,

1 Introduction

Bats are one of the most successful animals to populate the nocturnal niche and are amongst the most diverse orders of mammals, with close to 1,000 species (Altringham 1996). Part of their success in this niche is attributed to their ability to echolocate. Echolocation is a form of biological sonar in which the animal produces sounds and is able to detect objects based on the time taken for echoes to return (Neuweiler 2000). Bats, of the order Chiroptera, form two suborders, the Mega- and Microchiroptera. In fact, with one exception in the Megachiroptera, only the Microchiroptera possess echolocation. Nevertheless, this sensory system approaches state-of-the-art in these animals. Microchiropteran bats are able to use their echolocation system not only to orient in low or absent light levels but also to hunt and catch small insect prey. Echolocation performs at a level that makes the animal's ability to avoid obstacles indistinguishable from when using vision (Griffin et al. 1958). When hunting, bats are able to distinguish between objects and avoid capturing non-food items (Simmons et al. 1995). Echolocation call structure shows a remarkable diversity amongst bats, depending on the hunting strategy of the animal (Fenton 1995). Short-duration, frequency modulated pulses are generally used by animals hunting in open areas, allowing them to scan over a wider area and then narrow the search distance when a target is acquired. Constant frequency calls are used by animals hunting amongst foliage, allowing them to distinguish insect movement from background "clutter."

While the sophistication of echolocation in bats has reached its zenith in prey acquisition, it most likely evolved for spatial orientation (Schnitzler et al. 2003), allowing small volant animals to find their way around in the dark or in caves. Bats relying on echolocation for spatial memory appear to make stereotyped behaviours, often ignoring current sensory input once a memory of that area has been formed (Griffin 1958; Höller and Schmidt 1996; Neuweiler 1962). In the case of the experiment by Griffin, this tendency led to the disastrous consequence of the animals flying into a wall! Whilst their ability to learn a location in space appears unaffected by using echolocation rather than vision, the task is more difficult using the latter (Holland et al. 2005), requiring more time to learn (Fig. 1, pic of rosettus). Interestingly, the same area of the brain, the hippocampus, appears to be responsible for the spatial memory formed using echolocation as that formed using vision (Ulanovsky and Moss 2007). However, it seems that, if the memory is formed using vision, it does not transfer to when the bat is required to use echolocation for orientation; the task must be relearned (Holland et al. 2005).

2 Blind as a Bat?

The ability of bats to orient and catch prey using echolocation, along with apparently small eyes in many insectivorous bats, has led to the popular phrase "blind as a bat." Even in the scientific community. it is generally assumed that the enhanced

Fig. 1 The echolocating fruit bat, *Rousettus aegyptiacus*, seeks a hidden perch on one side of a two-alternative apparatus. This task is performed in visual conditions. When repeated in darkness, the bat has to relearn the task, indicating that despite the fact that both visual and echolocation spatial memory is processed in the hippocampus, the two do not interact (Holland et al. 2005)

auditory processing ability of echolocating bats must result in a trade-off elsewhere. However, "blind as a bat" turns out to be something of a misnomer. Not only do many insectivorous bats have good vision, comparable with other mammals of their body size (Eklöf 2003), but they actually prefer to use it in prey capture if it is available (Eklöf and Jones 2003), suggesting that while echolocation is a necessary component of the nocturnal lifestyle, it is not as effective a sensory mechanism as vision for these animals. This is not in fact surprising. Sound attenuates rapidly in air, meaning that the maximum range of echolocation is 20 m (Kick 1982). Even the recent discovery that bats can call as loud as 130 dB S.P.L. extends the range only by a small amount (Holderied et al. 2005; Holderied and von Helversen 2003). Bats are volant animals capable of moving large distances rapidly. Many bats make nightly foraging flights, and some bats migrate, making journeys of thousands of kilometers in the spring and autumn. A sensory mechanism that only allows a range of 20 m is clearly inadequate for orientation and navigation over large distances. How do bats, faced with the challenge of orientation and navigation over this range, manage to find their way home in the dark?

This chapter will review the mechanisms potentially available to bats to find their way around at night, based on current research on how night-migrating birds are able to orient and navigate. The theory of how animals orient and navigate and the potential sensory cues available will be discussed first, followed by the current knowledge of bat navigation. Finally new research that is shedding light on bat navigation will be discussed.

3 Theory of Orientation and Navigation

First proposed by Kramer (1953), the "map and compass" theory of animal navigation remains the most viable hypothesis today. This theory postulates that animals undergo a two-step process to navigate to a goal. The first step is to determine their position, the map step. This step gives the animal its location with respect to the desired goal. Having achieved this, the animal needs to take up the direction necessary to reach its goal, the compass step. Whether animals possess a map in the sense that humans think of a map, comprehensively covering the landscape between start and end points. is highly debated, but many animals are able to reorient themselves and return to a home site after being displaced many thousands of miles. This ability has been demonstrated most spectacularly in migratory birds.

Perdek (1958) discovered that, if adult migratory starlings were captured during their migratory journey from the Baltic to northern France and transported east to Switzerland, they were able to return to their normal wintering grounds. Juveniles, on the other hand, were unable to correct and carried on travelling in their normal migratory direction, ending up in Spain (Fig. 2a) A recent experiment on white crowned sparrows that were captured in Washington on the west coast of the United States and transported to New Jersey on the east coast indicated that adult birds were able to quickly correct for their displacement and were headed towards their normal wintering grounds in southern California by the time they were 25 km from the point of displacement (Thorup et al. 2007). As with Perdek's starlings, however, juveniles continued orienting south (Fig. 2b). These experiments tell us that adult migrating birds use a navigational map to locate their position, but juveniles rely on an inherited compass direction. In the case of the white crowned sparrows, this means that the navigational map must span the distance of the continental United States.

4 Sensory Cues for Orientation and Navigation

How do animals locate their position with respect to a goal and take up a direction to reach that goal? It seems that there are many environmental cues available to allow animals to achieve such feats. It has been proposed that, in order to locate their position, animals use a bi-coordinate system (Wallraff 1991) in which they learn the variation in intensity or concentration of two environmental factors that bisect each other, ideally at right angles (Fig. 3: bicoordinate). Theoretically, if they are able to learn that the two gradients vary predictably on a local scale, providing that those gradients continue to follow that pattern of variation on a larger scale (ideally

a

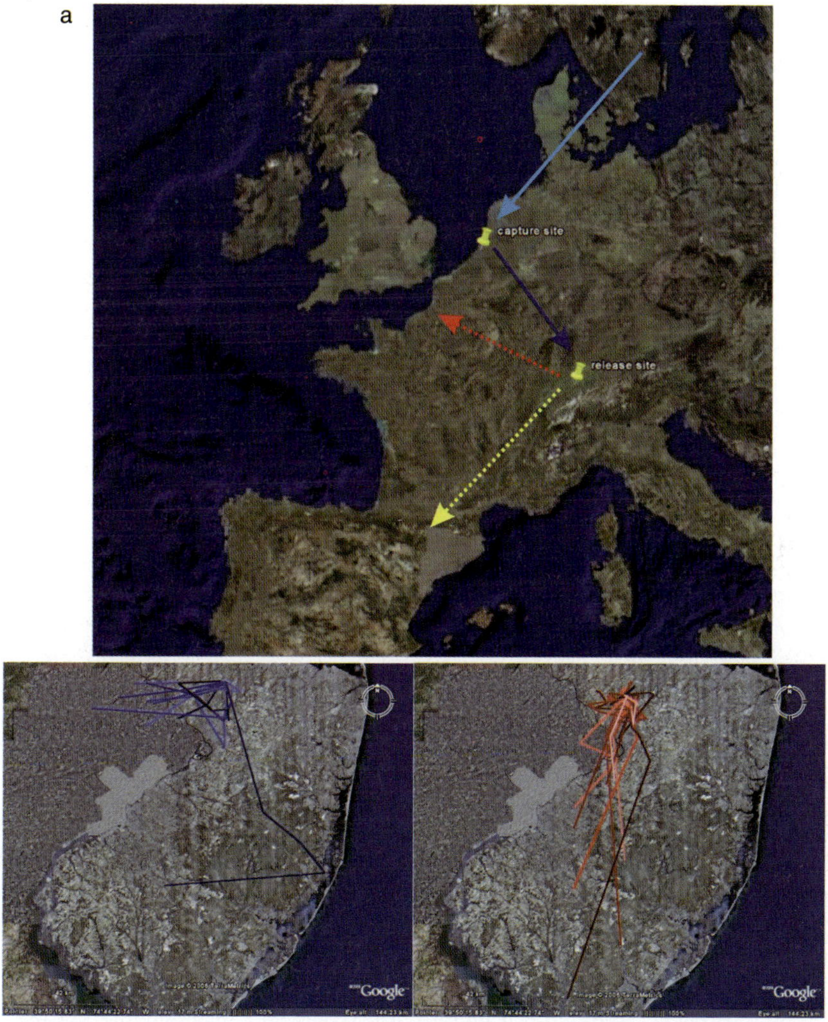

b

Fig. 2 **a**. The Classic displacement experiment (Perdek 1958). Birds migrating from the Baltic (light blue arrow) were caught in Holland and translocated to Switzerland (blue arrow). The location of ringing recoveries indicated that juveniles had continued migrating south west, (yellow dotted arrow), whereas adults had corrected and returned to known wintering grounds for the population (red dotted arrow). **b**. The paths of adult (blue tracks) and juvenile (red tracks) white-crowned sparrows *Zonotricia leucophrys* after being displaced to Princeton, New Jersey, from Seattle, Washington, in the USA. Adult birds head towards their species wintering grounds in California, whereas juvenile birds continue on a southerly heading, as if they were still on the west coast of the US (from Thorup et al. 2007)

globally), and if the animals are displaced to a location outside an area that is familiar, then all they need to do to determine their position is to compare the values of the two gradients with the known values at its goal to indicate their longitudinal and latitudinal displacement. Although no animal has as yet been clearly shown to

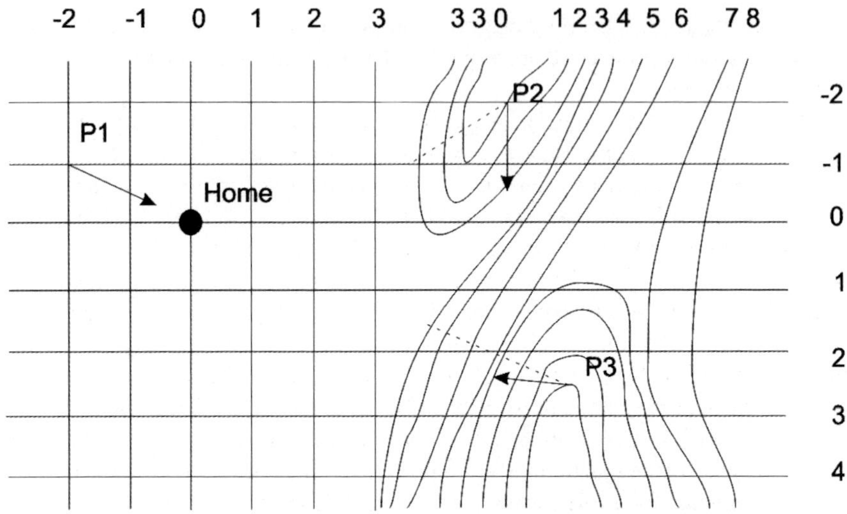

Fig. 3 A theoretical representation of a bicoordinate grid map. At P1, the animal compares the current values of the two gradients to the home site and is able to work out its relative position and, therefore, the direction needed to return home. Ideally the gradients will vary predictable on a global scale, but in practice this probably does not happen. At P2 and P3, the gradients are not perfect and so there are errors in the direction indicated compared to the actual direction needed to reach the home site

navigate using this system, much research effort has been expended in discovering the "mystery" of bird navigation (Baker 1976), in the main by discovering which senses are necessary for the animal to be able to successfully navigate back to a familiar goal. Early experiments tested the role of the sun's arc and altitude, which varies with latitude (Matthews 1970) and coriolis force (Yeagley 1947). Only two senses have stood the rigour of experimentation, however. It has been proposed that homing pigeons need an intact olfactory sense to navigate (Benvenuti et al. 1973; Papi 2001; Wallraff 2005), as demonstrated by a variety of methods, including sectioning of the olfactory nerve, nasal anaesthesia and washing the olfactory epithelia with zinc sulphate, along with a number of experiments manipulating the odour source to deflect the orientation of the pigeons (Wallraff 2005, for review). Exactly how olfactory cues are used in a gradient map has yet to be determined. It appears that exposure to odours brought in by winds at the home loft during development is an important factor in the ability to navigate by olfactory cues later in life (Ioale et al. 1990; Wiltschko and Wiltschko 1989), but it is unclear whether these form a bi-coordinate system.

Magnetic cues have been proposed as at least one component of a bi-coordinate system. The use of the Earth's magnetic field as a cue for navigation is highly appealing, as its variation in intensity with latitude potentially provides a means of determining position on a global scale (Freake et al. 2006). It has been argued that homing pigeons use the Earth's magnetic field for navigation (Dennis et al. 2007; Wiltschko and Wiltschko 2006). However, an experiment in which the trigeminal nerve was severed found no effect on the ability of homing pigeons to navigate to their home loft (Gagliardo et al. 2006). The trigeminal nerve has been identified as being responsible for signalling magnetic intensity in pigeons (Mora et al. 2004), so this would seem to indicate that magnetic intensity is not a component of the map of homing pigeons.

Other evidence for a role for magnetic cues in bird navigation comes from the results of pulse re-magnetisation experiments on migratory birds. The biogenic mineral magnetite has been found in animal cells. If animals use this magnetic mineral to detect the Earth's magnetic field (Kirschvink and Gould 1981), a strong, short-duration pulse applied to an animal should re-magnetize the magnetite, causing it to give erroneous information about the magnetic field. Several experiments on migratory birds in orientation cages have shown that a magnetic pulse deflects their orientation (Beason et al. 1995; Wiltschko et al. 1994). Interestingly, only adult birds are affected (Munro et al. 1997), not juveniles, suggesting that the effect is on an experience-based system, presumably the navigational map. Severing the trigeminal nerve stops the effects of the pulse on orientation (Beason and Semm 1996), suggesting that it is magnetic intensity that is being detected.

Other evidence for a role for magnetic intensity comes from simulated displacements in juvenile sea turtles. These animals show a homing response and. if they are placed in a tank surrounded by a Helmholtz coil that alters the intensity and inclination of the magnetic field to represent locations north or south of their home site, they swim in the direction that would be expected in order to reach home from the displaced location (Lohmann et al. 2004).

While magnetic intensity may indicate latitude, the way animals determine longitude in conjunction with this has not yet been determined. Potentially, the difference between true north and magnetic north, declination, could be used, but this has not yet been demonstrated to be the case. Interestingly, some migratory birds do have a mechanism to determine declination by calibrating the magnetic field with sunset (Cochran et al. 2004; Muheim et al. 2006b).

Sensory cues used for compass orientation are far better understood that those used for map navigation. The sun (Schmidt koenig 1960), the stars (Emlen 1967) and the Earth's magnetic field (Wiltschko and Wiltschko 1972) have all been demonstrated to be used by animals for taking up a direction, and even may be used to calibrate each other (Muheim et al. 2006a; Wiltschko and Wiltschko 1995). The most controversy in the role of compass cues has been over how animals can perceive the Earth's magnetic field, but this will be discussed in more detail in a later section. The preceding section has mainly focused on migrating birds, despite this chapter being about orientation and navigation in bats, because far less is known about how bats orient and navigate at night.

5 Orientation and Navigation in Bats: Known Unknowns or Unknown Unknowns?

In birds, two model systems have been established to study navigation. The directed migratory restlessness behaviour that migratory birds display if caged during their migratory period can be used as a baseline for testing navigation mechanisms.

Additionally, homing pigeons make an ideal model for field-based study of navigation, being domesticated and so being motivated to returning to a home loft. Bats have no such model species to test and so research on orientation in bats has been relatively rare compared to bird navigation. A recent paper proposed that bird navigation be used as a model to investigate the sensory cues and mechanisms used by bats for orientation (Holland 2007), and as technology has advanced there have been some recent exciting breakthroughs in the field. The remainder of this chapter will discuss what is known about bat navigation and report on the new findings.

Early research on bats suggested that they could home to a roost if displaced anywhere from 10 km to one extreme case of 450 km (Davis 1966). A number of studies investigating the role of vision in homing indicated that, if bats were blindfolded and displaced further than 12 km, they were unable to return to their roost. If displaced less than this, though, the bats were able to return, despite being blindfolded (Smith and Goodpaster 1958; Williams and Williams 1967, 1970; Williams et al. 1966). Further analysis indicated that if these animals were blindfolded and had their hearing blocked, they could not home (Stones and Branick 1969), suggesting that, despite echolocation having a range of only 20 m, bats were able to use it to navigate over distances of up to 12 km. How they achieve this is unknown and has not been investigated in any further detail, but it nevertheless seems that bats are able to build up a picture of the environment using echolocation that allows them to learn a familiar area from which they can return in the absence of visual cues.

Beyond the range of an echolocation-based, familiar area map, it seems that vision is essential for homing in bats. Once again, this fact emphasises the inaccuracy of the phrase "blind as a bat," but what does it tell us about bat navigation? The use of vision for homing might suggest navigation by familiar landmarks. Homing pigeons are known to use vision for homing from a familiar site (Biro et al. 2002). However, in some of the experiments, returns were noted from > 60 km, a distance that seems unlikely to be within the range of a familiar area. What cues could the animal use for navigation that would require vision at night? Clearly the sun is out of the question, but the stars may be used as a compass. One experiment suggested that bats do not have the visual acuity to use a stellar compass (Suthers and Wallis 1968), but a more recent study showed that bats were capable of discriminating between artificial light sources representative of stars (Childs and Buchler 1981). Given that the sun is out of the question and the magnetic field is the other potential source of compass information, one might assume that bats use a star compass to orient. It turns out. though, that some animals might perceive the magnetic field through their eyes (Wiltschko and Wiltschko 2006). A number of experiments have indicated that migratory birds cannot orient by a magnetic compass in the absence of certain wavelengths of light (Wiltschko and Wiltschko 2002). It also appears that this ability is lateralized to the right eye, as covering this eye also precludes birds from orienting by the magnetic compass (Wiltschko et al. 2002b). It has been proposed that this mechanism of detection of the magnetic field is based upon the change of state of photo-pigments in the eye in response to the Earth's magnetic field (Ritz et al. 2000). While it might be thought that orienting at night, as bats do, would preclude a magnetic compass mechanism based on light, it has also been demonstrated that salamanders use this mechanism in nocturnal conditions (Diego-Rasilla et al. 2005). On this basis, a navigation mechanism based on vision cannot preclude either stars or the magnetic field.

There has been no significant research on orientation and navigation in bats for 30 years, but recently new breakthroughs have been achieved that have shed light on the way these nocturnal animals orient. Holland et al. (2006) investigated the possibility that bats orient using a magnetic compass. They used a recent experiment that indicated that migratory thrushes calibrate the magnetic compass by the sunset as a model for their investigation (Fig. 4). Using big brown bats, *Eptesicus fuscus*, they discovered that these animals would rapidly return to their home roost if displaced 20 km north of it, providing a baseline to investigate the possible role of the stars and the Earth's magnetic field in their orientation. By exposing the bats to an altered magnetic field at sunset and until only stars were visible, it was possible to test whether bats used the Earth's magnetic field or the stars as a compass (Fig. 4 for predictions). The bats were deflected according to the prediction expected for the use of a magnetic compass calibrated by the sunset (Fig. 5), indicating that bats have a magnetic compass for orientation and, like birds, they calibrate it to the direction of sunset (Holland et al. 2006). This would allow them to correct for the difference between true north and magnetic north (declination).

How do bats perceive the magnetic field? As mentioned above, some animals such as birds and amphibians appear to "see" the magnetic field, but it has also been demonstrated that the mineral magnetite is used for magnetic field detection. First implicated in the orientation of magnetotactic bacteria (Blakemore 1975), a magnetite-based magnetoreceptor could provide polarity information in the same way that a handheld magnetic compass does if the magnetite were free to rotate within a cell, effectively acting like a tiny compass needle. In addition, it could provide information about the intensity of the magnetic field based on oscillations or torque around the magnetic moment (Kirschvink and Gould 1981). Figure 6 demonstrates in a simplified form how the magnetite could be used to detect the magnetic field. In birds, it has been proposed that the compass is detected by a light-dependent mechanism whereas magnetite is used to detect intensity for a magnetic map (Wiltschko and Wiltschko 2006). An alternative hypothesis suggests that magnetite-based magnetoreceptors alone are responsible for detecting the magnetic field both for compass and for map use (Kirschvink et al. 2001). Current evidence suggests that, for birds and salamanders at least, the former is correct. In the only mammal so far investigated, it appears that magnetite is used (Wegner et al. 2006). It has been argued that magnetoreception is divided along taxonomic lines (Wiltschko and Wiltschko 2006), but as both night-migrating birds and salamanders are able to use a light-dependent compass at night, it is possible that ecological factors play a role. A recent experiment using a magnetic pulse parallel and antiparallel to a north-south biasing field has indicated that big brown bats use magnetite to detect the Earth's magnetic field. By this technique, the polarity of the magnetite would be reversed in the antiparallel group but not the parallel group and in tests, only bats that received a pulse antiparallel to the magnetic field had orientation that was affected by the pulse (Holland et al. 2008). This finding is consistent with the discovery that

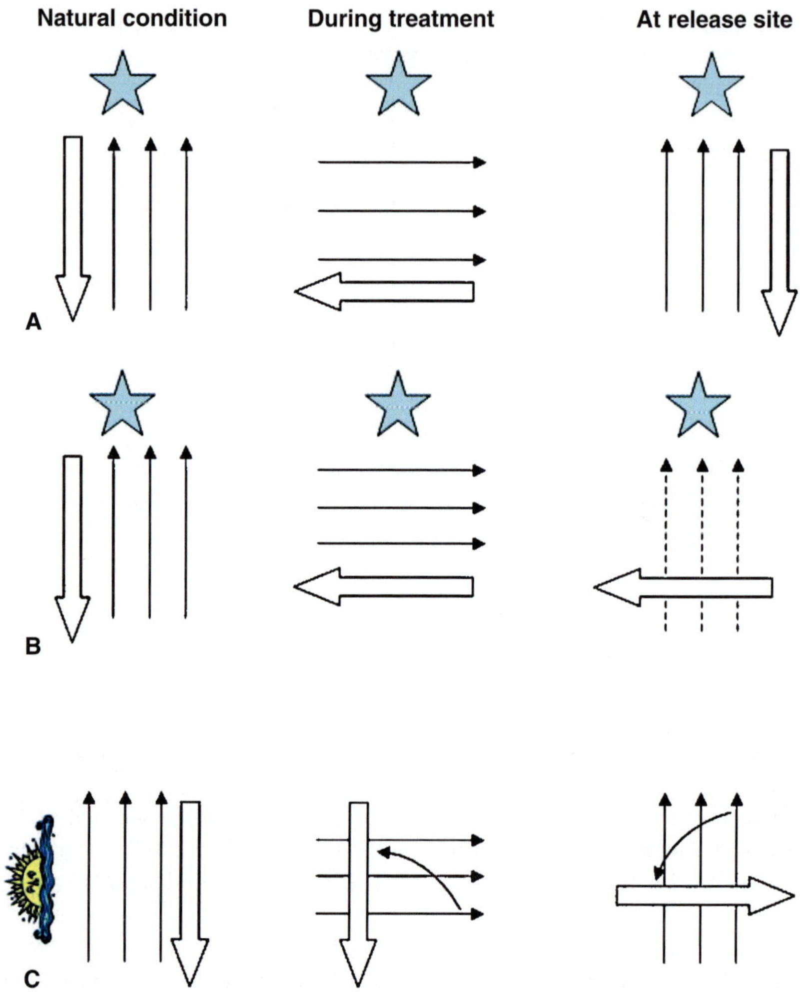

Fig. 4 Schematic diagram to illustrate the predictions for direction of orientation after exposure to a magnetic field shifted clockwise (CW) through sunset, depending on the interaction between magnetic and celestial cues. The three lined arrows in each diagram represent the magnetic field direction and the solid arrow represents the perceived direction of the home site for bats displaced north of home. **A**. If there is simple dominance of the magnetic compass or a star compass, with no calibration, then when released in a natural magnetic field at the release site the bats should fly south regardless of the direction of the field during treatment. **B**. If the magnetic field calibrates a star compass, then during treatment the field is shifted CW, making the bat perceive north as east and therefore south as west. When released, a bat trying to fly south to reach home would fly west. **C**. If sunset cues calibrate a magnetic compass, then when the field is shifted east at sunset the bat will perceive that geographical south is 90° CW to the magnetic field direction. When released in a normal magnetic field, flying 90° CW to it will result in the bat flying east (from Holland et al. 2006)

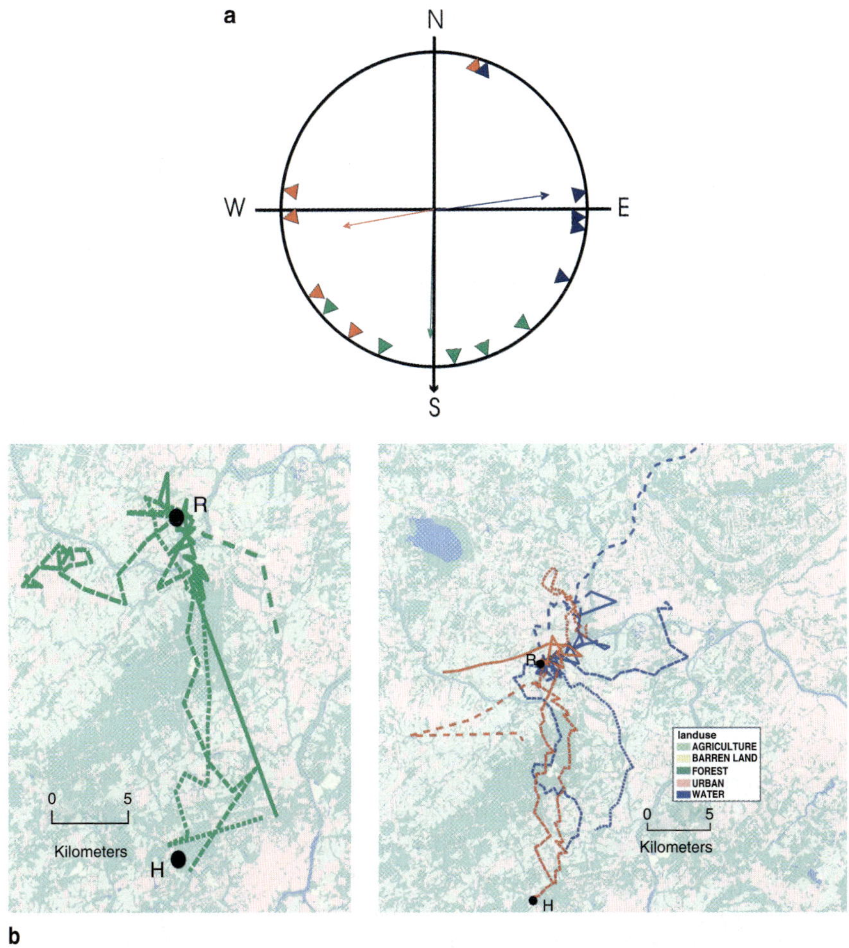

Fig. 5 a. Circular diagram of headings all groups released at the site 20 km north of home. Red=counter clockwise rotation of the magnetic field at sunset (CCW), blue=clockwise rotation of the magnetic field at sunset (CW) and green=control bats. The single-headed arrow outside the circle indicates the home direction (South). **b**. Control and experimental tracks of bats followed by radio telemetry, as reconstructed from GPS waypoints. Red=CCW magnetic field at sunset, blue=CW magnetic field at sunset and green=control. R=release site, H=home (from Holland et al. 2006)

the magnetic compass is based on the polarity of the magnetic field in other bats (Wang et al. 2007). The magnetic compass of birds appears to be insensitive to polarity, relying instead on detecting the inclination of the magnetic field, also known as the "dip" angle (Wiltschko and Wiltschko 1972). Exactly how the magnetite is able to signal within the cell the direction and intensity of the magnetic field is still theoretical (Kirschvink et al. 2001) and remains something of a black box. Bats have been shown to possess magnetic remanence, indicating the presence of magnetite in their bodies (Buchler and Wasilewski 1985), but exactly where the sensory

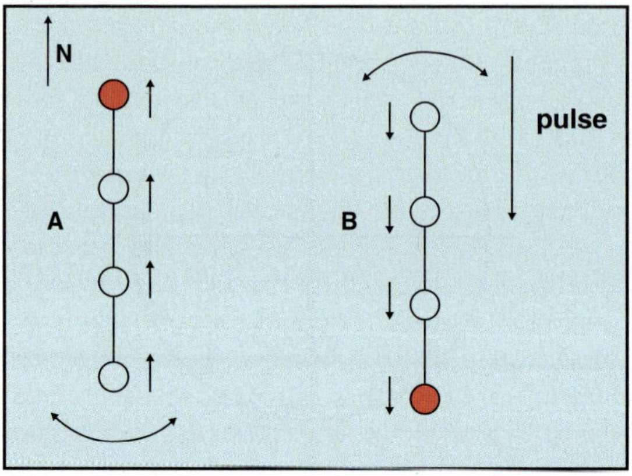

Fig. 6 A schematic representation of the theoretical behaviour of magnetite chains in a sensory cell if they are free to rotate. Normally the chain aligns to the Earth's magnetic field as in **A**. One end is "painted red" to indicate the direction of the magnetic field. How the cell signals this in practice is still unknown. The strength of oscillations around this direction is one possible way in which the intensity of the magnetic field might be measured. In **B**, when a strong, short-duration magnetic pulse is applied to the magnetite when aligned with a biasing field stronger than the earth's, if the magnetite is free to rotate it will flip its orientation and the "compass needle" will now be pointed south instead of north, causing the animal to reverse its orientation. In bats, recent evidence suggests that this is the case (Holland et al. 2008), whereas in birds, the results of pulse experiments do not support freely rotating magnetite (Wiltschko et al. 2002a)

cells containing magnetite are located is still unknown. How the bats, or indeed any animal, "perceive" the magnetic field using magnetite has yet to be demonstrated. In the mole rat, which uses a magnetite-based magnetic compass (Marhold et al. 1997), magnetite has been discovered in the eye, but whether this means they "see" the magnetic field is unclear (Wegner et al. 2006). In both the experiments on big brown bats, the animals were eventually able to correct for their faulty orientation and return home, suggesting that bats have other compass mechanisms that can be used for orientation. It remains unknown what cues if any are used by bats for map navigation. The presence of magnetite in their bodies gives them the potential to use the intensity of the magnetic field. Experiments on homing performance after nasal occlusion did not indicate any effect of the manipulation (Mueller 1963; Twente 1955), perhaps suggesting that olfactory cues do not play a role, but the experiments were performed at a distance from which bats have been shown to be able to home using echolocation, so no firm conclusions can be made about the role of olfactory cues.

6 How Bats Perceive the World

In this chapter we have seen that bats are capable of homing using echolocation from as much as 12 km from their roost, despite the fact that echolocation has a range of at most 20 m. Beyond this range it appears that vision is essential for homing.

Recently it has been shown that bats use magnetite to detect the Earth's magnetic field and use it as a compass, but whether vision is required for this is unknown. Potentially bats could call upon a number of sensory cues to locate their position when displaced outside a familiar area, but which cues are essential have yet to be determined. While we are starting to make new discoveries on how bats perceive the world to allow them to orient and navigate at night, there is still much to learn. How do bats string together 20-m bursts of information from echolocation to allow them to learn a familiar area 12-km wide? What cues do bats use to correct when the magnetic compass gives them misleading information? How is the magnetic field, perceived by magnetite in sensory cells, translated into information that the bat can use to take up a compass direction? How do they locate their position when outside the familiar area? After 30 years of inactivity in this field, we are finally starting to make inroads and there must surely be more exciting discoveries to come in the field of navigation at night.

References

Altringham JD (1996) Bats, biology and behaviour. Oxford University Press, Oxford

Baker RR (1976) Bird navigation: the solution of a mystery? Hodder Arnold, London

Beason RC, Semm P (1996) Does the avian ophthalmic nerve carry magnetic navigational information? J Exp Biol 199: 1241–1244

Beason RC, Dussourd N, Deutschlander ME (1995) Behavioral evidence for the use of magnetic material in magnetoreception by a migratory bird. J Exp Biol 198: 141–146

Benvenuti S, Fiaschi V, Fiore L, Papi F (1973) Homing performances of inexperienced and directionally trained pigeons subjected to olfactory nerve section. J Comp Physiol 83: 81–92

Biro D, Guilford T, Dell'Omo G, Lipp HP (2002) How the viewing of familiar landscapes prior to release allows pigeons to home faster: evidence from GPS tracking. J Exp Biol 205: 3833–3844

Blakemore RP (1975) Magnetotactic bacteria. Science 190: 377–379

Buchler ER, Wasilewski PJ (1985) Magnetic remanence in bats. In: Kirschvink JL, Jones DS, MacFadden BJ (eds) Magnetite biomineralization and magnetoreception in organisms: a new biomagnetism. Plenum Press, New York, pp 483–488

Childs SB, Buchler ER (1981) Perception of simulated stars by Eptesicus-fuscus (Vespertilionidae) - a potential navigational mechanism. Anim Behav 29: 1028–1035

Cochran WW, Mouritsen H, Wikelski M (2004) Migratory songbirds recalibrate their magnetic compass daily from twilight cues. Science 304: 405–408

Davis R (1966) Homing performance and homing ability in bats. EcolMon 36: 201–230

Dennis TE, Rayner MJ, Walker MM (2007) Evidence that pigeons orient to geomagnetic intensity during homing. Proc Roy Soc B-Biol Sci 274: 1153–1158

Diego-Rasilla FJ, Luengo RM, Phillips JB (2005) Magnetic compass mediates nocturnal homing by the alpine newt, Triturus alpestris. Behav Ecol Sociobiol 58 361–365

Eklöf J (2003) Vision in echolocating bats. Department of Zoology, Goteborg University, Goteborg, 104 pp

Eklöf J, Jones G (2003) Use of vision in prey detection by brown long-eared bats, *Plecotus auritus*. Anim Behav 66: 949–953

Emlen ST (1967) Migratory orientation in indigo bunting Passerina cyanea .I. Evidence for use of celestial cues. Auk 84: 309–318

Fenton MB (1995) Natural history and biosonar signals. In: Popper AN, Fay RR (eds) Hearing by bats. Springer-Verlag, New York, pp 37–86

Freake MJ, Muheim R, Phillips JB (2006) Magnetic maps in animals: A theory comes of age? Quart Rev Biol 81: 327–347

Gagliardo A, Ioale P, Savini M, Wild JM (2006) Having the nerve to home: trigeminal magnetoreceptor versus olfactory mediation of homing in pigeons. J Exp Biol 209: 2888–2892

Griffin DR (1958) Listening in the dark. Yale University Press, New Haven

Griffin DR, Novick A, Kornfield M (1958) The sensitivity of echolocation in the fruit bat *Rousettus*. Biol Bull 155: 107–113

Holderied MW, von Helversen O (2003) Echolocation range and wingbeat period match in aerial-hawking bats. Proc Roy Soc London: Series B 270: 2293–2299

Holderied MW, Korine C, Fenton MB, Parsons S, Robson S, Jones G (2005) Echolocation call intensity in the aerial hawking bat Eptesicus bottae (Vespertilionidae) studied using stereo videogrammetry. J Exp Biol 208: 1321–1327

Holland RA (2007) Orientation and navigation in bats: known unknowns or unknown unknowns? Behav Ecol Sociobiol 61: 653–660

Holland RA, Winter P, Waters DA (2005) Sensory systems and spatial memory in the fruit bat *Rousettus aegyptiacus*. Ethology 111: 715–729

Holland RA, Thorup K, Vonhof MJ, Cochran WW, Wikelski M (2006) Bat orientation using Earth's magnetic field. Nature 444: 653

Holland RA, Kirschvink JL, Doak T, Wikelski M (2008) Bats use magnetite to detect the Earth's magnetic field. PLoS One 3: e1676

Höller P, Schmidt U (1996) The orientation behaviour of the lesser spearnosed bat, *Phyllostomus discolor* (Chiroptera) in a model roost: concurence of visual, echoacoustical and endogenous spatial information. J Comp Physiol A 179: 245–254

Ioale P, Nozzolini M, Papi F (1990) Homing pigeons do extract directional information from olfactory stimuli. Behav Ecol Sociobiol 26: 301–305

Kick SA (1982) Target detection by the echolocating bat, *Eptesicus fuscus*. J Comp Physiol 145: 431–435

Kirschvink JL, Gould JLB (1981) Biogenetic magnetite as a basis for magnetic field detection in animals. BioSystems 13: 181–201

Kirschvink JL, Walker MM, Diebel CE (2001) Magnetite based magnetoreception. Curr Opinion Neurobiol 11: 462–467

Kramer G (1953) Wird die sonnenhohe bei der Heimfindeorientierung verwertet. J Ornithol 94: 201–219

Lohmann KJ, Lohmann CMF, Ehrhart LM, Bagley DA, Swing T (2004) Geomagnetic map used in sea turtle navigation. Nature 428: 909–910

Marhold S, Burda H, Kreilos I, Wiltschko W (1997) Magnetic orientation in common molerats from Zambia. Paper No 5. Orientation and navigation: birds, humans and other animals. Royal Institute of Navigation, Oxford

Matthews GV (1970) Do pigeons determine lattitudinal displacement from the suns altitude? Nature 227: 627–629

Mora CV, Davison M, Wild JM, Walker MM (2004) Magnetoreception and its trigeminal mediation in the homing pigeon. Nature 432: 508–511

Mueller HC (1963) Homing and distance orientation in bats. PhD, University of Wisconsin, Madison

Muheim R, Moore FR, Phillips JB (2006a) Calibration of magnetic and celestial compass cues in migratory birds - a review of cue-conflict experiments. J Exp Biol 209: 2–17

Muheim R, Phillips JB, Akesson S (2006b) Polarized light cues underlie compass calibration in migratory songbirds. Science 313: 837–839

Munro U, Munro JA, Phillips JB, Wiltschko R, Wiltschko W (1997) Evidence for a magnetite-based navigational 'map' in birds. Naturwissenschaften 84: 26–28

Neuweiler G (1962) Bau und Leistung des Flughundauges. Zeit Vergleichende Physiol 46: 13–56

Neuweiler G (2000) The biology of bats. Oxford University Press, New York

Papi F (2001) Animal navigation at the end of the century: a retrospect and a look forward. Ital J Zool 68: 171–180

Perdek AC (1958) Two types of orientation in migrating *Sturnus vulgaris* and *Fringilla coelebs*. Ardea 46

Ritz T, Adem S, Schulten K (2000) A model for vision based magnetoreception in birds. Biophys J 78: 707–718

Schmidt koenig K (1960) Internal clocks and homing. Cold Spring Harbor Symp Quant Biol 25: 389–393

Schnitzler HU, Moss CF,Denzinger A (2003) From spatial orientation to food acquisition in echolocating bats. Trends Ecol Evol 18: 386–394

Simmons JA, Ferragamo MJ, Saillant PA, Haresign T, Wotton JM, Dear SP, Lee DN (1995) Auditory dimensions of acoustic images in echolocation. In: Popper AN, Fay RR (eds) Hearing by bats.Springer-Verlag, London, pp 146–190

Smith E, Goodpaster W (1958) Homing in nonmigratory bats. Science 127: 644–644

Stones RC, Branick LP (1969) Use of hearing in homing by 2 species of Myotis bats. J Mammal 50: 157–160

Suthers RA, Wallis NE (1968) Optics of eyes of echolocating bats. Am Zool 8: 774–780

Thorup K, Bisson I, Bowlin M, Holland RA, Ramenofski M, Wingfield J, Wikelski M (2007) Migration routes of adult and juvenile white-crowned sparrows differ after continent-wide displacement during migration. PNAS 104: 18115–18119

Twente JW (1955) Some aspects of habitat selection and other behavior of cavern-dwelling bats. Ecology 36: 706–732

Ulanovsky N, Moss CF (2007) Hippocampal cellular and network activity in freely moving echolocating bats. Nature Neurosci 10: 224–233

Wallraff HG (1991) Conceptual approaches to avian navigation. In: Berthold P (ed) Orientation in birds Birkhauser Verlag, Berlin, pp 16–37

Wallraff HG (2005) Avian navigation: pigeon homing as a paradigm. Springer, Berlin

Wang Y, Pan Y, Parsons S, Walker MM, Zhang S (2007) Bats respond to polarity of a magnetic field. Proc Roy Soc B-Biol Sci 274: 2901–2905

Wegner RE, Begall S, Burda H (2006) Magnetic compass in the cornea: local anaesthesia impairs orientation in a mammal. J Exp Biol 209: 4747–4750

Williams TC,Williams JM (1967) Radio tracking of homing bats. Science 155: 1435–1436

Williams TC, Williams JM (1970) Radio tracking of homing and feeding flights of a neotropical bat, Phyllostomus-hastatus. Anim Behav 18: 302–309

Williams TC, Williams JM, Griffin DR (1966) Visual orientation in homing bats. Science 152: 677–679

Wiltschko W, Wiltschko R (1972) Magnetic compass of European robins. Science 176: 62–64

Wiltschko R, Wiltschko W (1989) Pigeon homing-olfactory orientation-a paradox. Behav Ecol Sociobiol 24: 163–173

Wiltschko R, Wiltschko W (1995) Magnetic orientation in animals. Springer, Berlin

Wiltschko W, Wiltschko R (2002) Magnetic compass orientation in birds and its physiological basis. Naturwissenschaften 89: 445–452

Wiltschko R, Wiltschko W (2006) Magnetoreception. Bioessays 28: 157–168

Wiltschko W, Munro U, Beason RC, Ford H, Wiltschko R (1994) A magnetic pulse leads to a temporary deflection in the orientation of migratory birds. Experientia 50: 697–700

Wiltschko W, Munro U, Wiltschko R, Kirschvink JL (2002a) Magnetite-based magnetoreception in birds: the effect of a biasing field and a pulse on migratory behavior. J Exp Biol 205: 3031–3037

Wiltschko W, Traudt J, Gunturkun O, Prior H, Wiltschko R (2002b) Lateralisation of magnetic compass orientation in a migratory bird. Nature 419: 467–470

Yeagley HL (1947) A preliminary study of a physical basis of bird navigation. J Appl Phys 18: 1035–1063

Point, Line and Counterpoint: From Environment to Fluid Space

Tim Ingold

Abstract This paper aims to understand what is meant by the environment of an animal and, more particularly, that of a human being. To avoid the contradictions entailed in assuming that human environmental relations are mediated by systems of symbolic meaning, with the absurd corollary that non-human animals inhabit meaningless worlds, I consider the sources of environmental meaning for non-humans and their possible availability to humans as well. In psychology, Gibson's theory of affordances offers one possible approach, though it is ultimately found to privilege the environment as a site of meaning vis-à-vis its inhabitants, whether human or non-human. In ethology, von Uexküll's theory of the Umwelt suggests, quite to the contrary, that meaning is bestowed by the organism on its environment. In philosophy, and following von Uexküll's lead, Heidegger drew a sharp distinction between the animal's "captivation" in its Umwelt and the way the world is disclosed, or opened up, to human beings. But the animal's captivation also implies a sense of openness, in the manner in which its life flows along lines comparable – in von Uexküll's terms – to those of polyphonic music. This sense has been taken up in the philosophy of Deleuze. The living organism, for Deleuze, is a bundle of lines, a haecceity. Critically, these lines do not connect points but pass forever amidst and between. Considering the way in which this idea has been taken up in actor-network theory, particularly associated with the work of Latour, I stress the importance of distinguishing the network as a set of interconnected points from the meshwork as an interweaving of lines. Every such line describes a flow of material substance in a space that is topologically fluid. I conclude that the organism (animal or human) should be understood not as a bounded entity surrounded by an environment but as an unbounded entanglement of lines in fluid space.

T. Ingold
School of Social Science, University of Aberdeen, Aberdeen AB24 3QY, Scotland, UK
e-mail: tim.ingold@abdn.ac.uk

A. Berthoz and Y. Christen (eds.), *Neurobiology of "Umwelt": How Living Beings Perceive the World,* Research and Perspectives in Neurosciences,

1 Beginning with the Environment

This paper is the latest chapter in my attempts over two decades and more - and that are still ongoing - to figure out what is meant by the environment of an animal. Coming from a background in ecological anthropology, which professes to study the relations between people and their environments, I cannot avoid the questions of what an environment is and, more particularly, what if anything is special about the environments of those animals we call human beings. Initially, my inquiries were prompted by a realisation that ecological anthropology appeared to have reached an impasse which was blocking further development in the subject. It lay in the contradictory imperatives, epitomised in the title of a celebrated book by Marshall Sahlins, of *Culture and Practical Reason* (Sahlins 1976). Does all meaning and value lie in systems of significant symbols? If so, then the motives and finalities for all human action on the environment must lie in what the mind brings to it: in the ideas, concepts and categories of a received cultural tradition. Yet does not culture with its artefacts and organisational arrangements, and the knowledge of how to apply them, provide human beings with the equipment to draw a livelihood from the world around them? Would they not, as Clifford Geertz once remarked (1973, pp. 49–50), be crippled without it? If so, then whence come the ultimate requirements of human practice if not from the environment itself? Precisely where are we to place culture in the nexus of human environmental relations? Does it dictate the terms of adaptation, or is it a means of adaptation on terms dictated by nature, or both at once?

All sorts of ingenious solutions had been proposed to this dilemma, branded with a bewildering array of cumbersome labels – cultural materialism, neofunctionalism, symbolic ecology, structural Marxism – whose very clumsiness was symptomatic of epistemological collapse. None of them offered a satisfactory way out. Searching around for an alternative approach, I began to wonder whether the source of the difficulty might lie in the one assumption that everyone had taken for granted, namely, that human relations with the environment are necessarily mediated by culture (Ingold 1992). After all, non-human animals that – with one or two possible exceptions – are not supposed to share the human capacity for symbolic representation are nevertheless quite well able to get along in their environments. Are we really meant to believe, as advocates of cultural reason would have it, that all meaning is symbolic, and therefore that non-humans inhabit meaningless worlds? To my mind, such a conclusion seemed absurd. So, to turn the question around, I asked: "What kind of meaning can there be in the absence of symbolic representation?" If we could just identify the sources of environmental meaning for non-human animals, then we could go on to consider the extent to which such sources are available to human beings as well. Only when these sources were exhausted would we finally need to resort to the sphere of cultural representation.

Looking for answers to my question, I found none in mainstream psychology or in the ethological study of animal behaviour. For the most part, cognitive psychologists were convinced that there could be no action in the world that was not preceded and determined in its course by an interior mental representation, that is, by an intention conceived in thought. If animals could not think or intend, then neither could they act. All they could do was behave, responding more or less automatically to received stimuli through innate mechanisms loosely known as "instincts." No meaning there! The majority of students of animal behaviour took the same view. Admittedly, there were mavericks such as Donald Griffin (1984), who surmised that even the lowly insects might be capable of deliberating over the course of action. They too assumed, however, that there could be no action without forethought. Their theory of meaning, which rested on a Cartesian split between the thinking mind and the executive body, diverged not at all from the mainstream; they differed only in where they drew the line, in the animal kingdom, between creatures with minds and creatures without. Yet is it not ironic that we should expect of the ant or bee, as a condition of its finding some meaning in the environment, that it holds before its mind some representation of the world and acts in accordance with it, when this is something we humans so rarely do ourselves? How often, I wonder, do we think before we act? Even when we do, the action hardly follows automatically from the thought, and may often diverge from it in ways never intended. As the philosopher Alfred North Whitehead wisely observed, "from the moment of birth we are immersed in action, and can only fitfully guide it by taking thought" (Whitehead 1938, p. 217).

I therefore had to leave the mainstream to find my answers. In psychology I turned to the work of James Gibson, whose ecological approach to perception, developed in 1950s and 1960s, was explicitly opposed to the prevailing paradigm of cognitivism. And in ethology I rediscovered the long neglected, pre-war writings of the Estonian-born pioneer of bio-semiotics, Jakob von Uexküll. Both seemed to offer a radically alternative way of thinking about meaning, finding it not in the correspondence between an external world and its interior representation but in the immediate coupling of perception and action. Yet, as I also found, behind this commonality lay significant differences.

2 James Gibson and the Concept of Affordance

Gibson's first move was to distinguish very clearly between "the animal environment" and the "physical world" (Gibson 1986, p. 8). Physics may strive to comprehend the nature of the world as it really is, pared down to its essential constituents of force, energy and matter. An environment, however, does not exist in and of itself. It exists only *in relation* to the being whose environment it is. Thus, just as there can be no organism without an environment, so also there can be no environment without an organism (see also Lewontin 1982, p. 160). Though no less real than the physical world, the environment is reality *for* the organism in question (Ingold 1992, p. 44; 2000, p. 168). Gibson's next step was to show that the fundamental constituents of any environment comprised what he called *affordances* (Gibson 1986, p. 127). His argument was that in encountering any particular environmental object, the animal perceived what it facilitated or hindered in the immediate context of its current activity. Perception, then, was not a matter of affixing some meaning to the object – of recognising it as one of a certain kind to which certain uses might be attached – but of discovering meaning in the very process of use.

Despite the clarity of Gibson's reasoning, it is in fact shot through with contradiction. The problem lies in his inability to reconcile his relational understanding of the environment with an older and more conventional view that posits the environment as a set of objective conditions that exist independently and in advance of the creatures that come to inhabit it, and to which they must perforce adapt. His solution is to try to have it both ways, as the following passage reveals:

> "An important fact about the affordances of the environment is that they are in a sense objective, real, and physical, unlike values and meanings, which are often supposed to be subjective, phenomenal and mental. But, actually, an affordance is neither an objective property nor a subjective property; or it is both if you like. An affordance cuts across the dichotomy of subjective-objective and helps us to understand its inadequacy. It is equally a fact of the environment and a fact of behavior. It is both physical and psychical, yet neither. An affordance points both ways, to the environment and to the observer" (Gibson 1986, p. 129).

Are affordances, then, objectively and physically instantiated in the environment prior to the assignation to them of value and meaning by a perceiving subject? As a matter of fact they are, says Gibson, before immediately qualifying himself. Well, they are "in a sense." And *actually*, he goes on to say, that sense rests on the entirely inadequate foundation of a subject-object dualism! For the affordances of things *are* their values and meanings, and what is more, they can be directly perceived (Gibson 1986, p, 127).

I believe the root source of this contradiction can be found in the very assumption that the environment comprises a world furnished with *objects*. For Gibson this is axiomatic. Imagine a world devoid of objects, with nothing but a bare, featureless ground stretching to the great circle of the horizon under a cloudless sky. What a desolate place it would be! We could stand and walk in it, but little else. It is the furniture of the earth, Gibson surmised, that makes it habitable (Gibson 1986, p. 78). In practice, inhabitants find themselves in a world cluttered with objects of all sorts, like householders in an attic or actors on a stage-set. From this analogy is drawn the classical ecological concept of the *niche*, a little corner of the world into which an organism has fitted itself through a process of adaptation. Just as, literally, an alcove in the wall provides the perfect place to display a vase of the right size and proportion, so metaphorically, every kind of creature has evolved to fill its particular niche in the environment. A corollary of the metaphor, however, is that as with the dimensions of the alcove, the niche is specified by essential properties of the environment, irrespective of the presence and functioning of the organism. Take away the vase, and the alcove is still there; remove the organism and the niche remains. As "a set of affordances" (Gibson 1986, p. 128), the niche is already laid out in the furnishing of the environment before any creature arrives to fill it. It sets the conditions to which any occupant must adapt. Moreover, every object of furniture, Gibson insisted, "offers what it does because of what it is" (Gibson 1986, p. 139), whether or not any animal is present to detect it. As properties of the furnished world, the affordances of the environment are there to be discovered and put to use by any creature with the equipment to do so.

In short, far from inhering in a relation between a living being and its environment, and pointing both ways, it now seems that the affordance rests unequivocally on the side of the environment and that it points in just one way, towards any potential inhabitant. Having begun by assuring us that "an environment implies an animal (or at least an organism) to be surrounded," Gibson went on to assert, with equal assurance but quite to the contrary, that "the environment does *not* depend on the organism for its existence" (Gibson 1986, pp. 8, 129; my emphasis). Indeed he was at pains to distinguish his view of the niche from "what some animal psychologists have called the *phenomenal environment* of the species," and particularly from any suggestion that such an environment might amount to a "subjective world" in which it is supposed to live (Gibson 1986, p. 129). Though he does not name names, he could have been referring to the works, among others, of Jakob von Uexküll.

3 Jakob von Uexküll and the Concept of Umwelt

Much as Gibson was later to do, von Uexküll set out to understand how the world exists *for* the animal, given its own particular morphology, sensibilities and action potentials. No more than Gibson could he accept that animals live in meaningless worlds. One could hardly imagine an animal farther removed from human beings in structure, size and complexity – though not, irritatingly, in proximity – than the humble tick. Yet even for the tick, von Uexküll showed, the environment is imbued with meaning, albeit of only three kinds (von Uexküll 1992, pp. 324–325). The first is carried in the smell of sweat common to mammals, the second in characteristics of the host's skin and hair, and the third in the temperature of warm blood. The significance of each lies in the action it prompts: falling (so as to land on the host), burrowing (on a relatively hairless patch of skin) and sucking (from blood vessels close to the surface). For von Uexküll as for Gibson, there is meaning in the animal's world not because it is capable of fashioning an internal representation of an external state of affairs but because its action in the world is so closely and intimately attuned to its perception (von Uexküll 1992, p. 320).

That is where the similarity ends, however. For whereas Gibsonian affordances are supposed to exist as the inherent potentials of environmental objects, regardless of whether they are attended to or put to use by any organism, von Uexküll maintained that what he called the "quality" (*Ton*) of a thing, by virtue of which it has significance for a particular creature, is not intrinsic to the thing itself but is *acquired* by virtue of its having been drawn into that creature's activity (von Uexküll 1982, pp. 27–29). The same stone, for example, may function as shelter for the crab that hides beneath it, as an anvil for the thrush that uses it to break open snail shells, and as a missile for an angry human to hurl at an adversary. In Gibson's terms, shelter, anvil and missile are all properties of the stone that are available to be *taken up*. For von Uexküll, by contrast, they are qualities that are *bestowed upon* the stone by the need of the creature in question and in the very act of attending to it. The stone only *becomes* a shelter when the crab scuttles under it, an anvil when the thrush smashes the shell against it, and a missile when the man picks it up to throw. Outside of these activities, it was none of these things. Thus, far from fitting into a given corner of the world (a niche), it is the animal that fits the world to itself by ascribing functional qualities to the things it encounters and thereby integrating them into a coherent system of its own (von Uexküll 1992, pp. 360–361; see Ingold 1992, p. 42). To denote this system – the world as it is constituted within the animal's circuit of perception and action – von Uexküll used the term *Umwelt* (1992, p. 320). The life of every creature, von Uexküll thought, was so wrapped up in its own Umwelt that no other worlds were accessible to it. It is as though each one were floating in its own particular "bubble" of reality (von Uexküll 1992, pp. 338–339). Though the perceptual and effector organs of different creatures may be perfectly attuned, neither can access what is real for the other. For example, the threads of the spider's web, as von Uexküll elegantly showed (1982, p. 42), are precisely proportioned such that they evade the visual sensors of the fly, yet the spider knows absolutely nothing of the fly's world.

We have seen that the niche, as a set of affordances, is on the side of the environment and points towards the organism. The Umwelt, it now seems, is just the opposite: it is on the side of the organism pointing towards the environment. Remove the organism, and the Umwelt disappears with it. What then remains? A man may throw a stone in anger, but in more measured circumstances he might ponder its possible uses as a paperweight, pendulum bob or hammer. Whilst he holds the stone in his hand and deliberates on the matter, it is not yet any of these things. It is merely an object of a certain shape, size and composition, with certain properties of hardness and durability, which could, in principle, find an almost unlimited range of uses. Regarded as such, the stone is an example of what von Uexküll (1982: 27) called "neutral objects." No animal, however, or at least no *non-human* animal, is in a position to observe the environment from such a standpoint of neutrality. To live, it must already be immersed in its surroundings and committed to the relationships this entails. And in these relationships, the neutrality of objects is inevitably compromised. The thrush, for example, does not first perceive the stone *as* a stone, and then wonder what to do with it, any more than it wonders what to do with its beak. Rather, using both stone and beak, it smashes shells. But what of the human? In a paper published almost twenty years ago, I argued that humans are different. Uniquely among animals, it seemed to me, human beings are capable of making their own life activity the object of their attention, and thus of seeing things *as they are*, as a condition for deliberating about the alternative uses to which they might be put (Ingold 1989, pp. 504–505). For this reason I took exception to the conventional English translation of the German Umwelt as "subjective universe" (e.g., von Uexküll 1982, p. 31). For human beings alone, I thought, can exist as subjects confronting a world of neutral objects. In that very act of standing back and reflecting on the conditions of existence, the human Umwelt becomes an *Innenwelt* – literally, a "subjective universe" – an organisation of representations, internal to the mind, which lend meaning to the raw material of experience.

4 Martin Heidegger on Life in the Open

In retrospect, I am no longer so sure that such a radical contrast can be drawn between human and non-human perceptions of the environment. For in the ordinary course of life, it now seems to me, humans no more have to reconstruct the world in the imagination – in the Innenwelt – as a condition for their practical engagement with it than do non-human animals. Much of the inspiration for this change of heart came from the philosophy of Martin Heidegger, though somewhat refracted through my reading of Hubert Dreyfus (1991, pp. 109–127). Heidegger distinguished between two ways in which things in the world can show up to a being active within it: *availableness* and *occurrentness*. To the skilled practitioner absorbed in an activity, the things he uses are available, ready-to-hand. So long as the activity flows smoothly, their objectness melts into the flow. As the practitioner's awareness becomes one with the activity, he or she does not attend to the objects as such. Hammering, the carpenter does not inspect the hammer; fiddling, the musician does not subject the violin to scrutiny. Only when the instrument fails to respond to the demands of the moment does the practitioner run hard up against it, in its brute facticity. In Heidegger's terms the thing is no longer available but occurrent. "What is this?" curses the carpenter as the hammer misses its mark, or the musician when the violin goes out of tune or a string snaps. This is not the kind of question that a non-human, without the gift of language, would ever ask. In this sense, humans alone are haunted by the spectre of the *loss* of meaning that occurs when action fails. It is not in their construction of meaningful worlds, then, that the singularity of human beings resides but rather in their occasional glimpses of a world rendered meaningless by its dissociation from action.

Should we conclude that nothing really distinguishes the perception of the animal in its Umwelt from that of the human practitioner, at least for so long as the latter is absorbed in the task at hand? This was certainly the drift of my own thinking, but it was not so for Heidegger. In a course of lectures delivered in 1929–30, but which lay unpublished until 1983, Heidegger set out his unequivocal stance on the question of human uniqueness in direct response to the work of von Uexküll, which he much admired. The animal in its Umwelt, he argued, may be open to its *environment*, but it is closed to the *world*. The human practitioner is unique in inhabiting the world of the open. To explain what he meant, Heidegger asked his listeners to compare an inanimate object like a stone, an animal and a human being. How do they differ? His answer took the form of three theses: "The stone … is *worldless*; the animal is *poor in world*; man is *world-forming*" (Heidegger 1995, p. 263). The stone has no world since it lacks a perceptual apparatus. Suppose that we find a stone lying on our path. "The stone lies upon the earth," observed Heidegger, "but does not touch it." Though it crops up amidst a host of other things, everything around remains inaccessible to the stone itself (Heidegger 1995, p. 197). There is, in short, no reality *for* the stone. What, then, of the animal? Why should its world have the character of poverty? If it is by the loss of meaning, and not by its contribution, that humans distinguish themselves from animals, then why are human worlds nevertheless more richly endowed?

The world of the animal is poor, Heidegger argued, because it is *captivated* (Heidegger 1995, p. 239). But as Giorgio Agamben has shown through a detailed commentary on Heidegger's text, there are two sides to captivation (Agamben 2004, pp. 49–56). On the one hand, although the animal is encircled within what Heidegger called a "disinhibiting ring," precisely equivalent to the Umwelt, this encirclement is absolutely not an encapsulation (Heidegger 1995, pp. 255, 263). For it is thanks to its ring of disinhibitors that the instinctual drives of the animal can be released and find expression in the presence of appropriate stimuli. The disinhibiting ring is like a ring of keys, each of which opens a door through which the life of the animal spills out into its surroundings. But the animal knows nothing of this. It completely fails to apprehend the things with which its life is mingled, *as things*. For the animal, driven to behave in the way it does, there is no possibility of apprehension (Heidegger 1995, p. 247). Thus the very same encircling ring that opens the animal to its environment also ensures that the world as we humans know it – infinitely extendable in range and possibility – is forever withheld from it (Heidegger 1995, p.193). This is the other side of captivation. The animal is poor in world, for Heidegger, because it lacks access to the things and beings that comprise it.

Yet if the closure entailed in the animal's captivation implies an openness to its environment, so conversely, the world of the human practitioner can be open only because it can appear closed in a way that the animal's world can never do. Since the world cannot be disclosed to the animal, there is no possibility, either, of its being *closed off* (Heidegger 1995, p. 248). For human beings, by contrast, the very opening of the world, the *dis*closure of things for what they are, is predicated upon an initial closure. Unlike the animal in its captivation, which finds itself *taken* in an environmental embrace that is as passionate as it is overwhelming, the human being stands before the world, as a domain of things-in-themselves, and has of necessity to take a stance *towards* it. Here, concludes Heidegger,

> "we see … the essential contrast between the animal's being open and the *world-openness* of man. Man's being open is a being held toward …, whereas the animal's being open is a being taken by … and thereby a being absorbed in its encircling ring" (Heidegger 1995, p. 343).

The contrast between these contrary understandings of openness and closure is epitomised in what Heidegger has to say, elsewhere, about hands and handicraft. No animal, he thinks, can have a hand or be handy. Animals can have paws, claws and talons, but these are mere conduits for its behaviour. The hand, by contrast, is an instrument of world forming. It is a hand precisely because it is *not* tied to any particular way of working, but delivers an engagement that is both thoughtful and reflexive, guided by consideration. It is, in short, an instrument not of *behaviour* but of *comportment* (Elden 2006, p. 280; see also Heidegger 1995, p. 237).

The peculiar boundedness of Heidegger's notion of the "open" is evident in his recurrent metaphor of the clearing, imagined as a space for dwelling that is *opened up* (that is, disclosed) from the surrounding forest. Within this space, human existence is reined in and contained, whereas other creatures meld into the surroundings from which they are deemed incapable of distinguishing themselves and to which they are therefore unable to relate *as such* (Agamben 2004, p. 59; Harrison 2007, p. 634). To be sure, Heidegger is anxious to avoid placing any hierarchical evaluation on the difference between the animal's poverty in the world and the human capacity for world-formation (Heidegger 1995, p. 194). That he should characterise the world in terms of what the human *possesses* and the animal *lacks* reveals, nevertheless, where his priorities lie. Poor animals (Elden 2006, p. 274)! Indeed, in his stress on human uniqueness, Heidegger seems to arrive at a picture of the inhabitant that is, in every respect, the precise inverse of Gibson's. Recall Gibson's contention that what he calls the *open environment* – realised in the limiting case as a perfectly level desert stretching to the horizon under an empty sky – would be practically uninhabitable (Gibson 1986, p. 33, 78). To create a space for dwelling, the open must be furnished with objects. Yet these objects, affording what they do because of what they are, remain indifferent to the presence of the inhabitant. They are supposed to comprise, in themselves, a meaningful world, into which the inhabitant arrives as a kind of interloper, probing this niche and that and picking up their affordances (Gibson, p. 139). For Heidegger, to the contrary, the space of dwelling is one that the inhabitant has formed around himself by clearing the clutter that would otherwise threaten to overwhelm his existence. The world is rendered habitable not as it is for Gibson, by its partial *en*closure in the form of a niche, but by its partial *dis*closure in the form of a clearing.

5 Gille Deleuze and Life on the Line

Can there be any escape from this shuttling back and forth between enclosure and disclosure, between an ecology of the real and a phenomenology of experience? So long as we suppose that life is fully encompassed in the relations *between* one thing and another – between the animal and its environment or the being and its world – we are bound to have to begin with a separation, siding either with the environment vis-à-vis its inhabitants or with the being vis-à-vis its world. A more radical alternative, however, would be to reverse Heidegger's priorities, that is, to celebrate the openness inherent in the animal's very captivation by its environment. This is the openness of a life that *will not be contained*, that overflows any boundaries that might be thrown around it, threading its way like the roots and runners of a rhizome through whatever cracks and crevices afford growth and movement. Once again, we can take our cue from von Uexküll, who compares the world of nature to polyphonic music, in which the life of every creature is equivalent to a melody in counterpoint (von Uexküll 1982, pp. 52–54). In the case of musical performance, we may speak of the connection between the player and his instrument, say a violin. Each has a bearing on the other. But the line of the melody does not lie in this connection. On the contrary, it is a line that continually issues forth from that place, in the midst of things, where the fiddler and the violin are conjoined in a passionate embrace. So too, the life-lines of organisms issue from the sites of their symbiotic connection, but in a direction that runs not from one to the other but forever in between, as the river flows between its banks in a direction orthogonal to their transverse connection. The life of the spider thus runs in counterpoint to that of the fly: to the melodic line of the first, the second figures as a refrain (von Uexküll 1982, p. 68). To adopt this view is to go with the grain of another of the twentieth century's most influential philosophers, Gilles Deleuze.

Life, for Deleuze, is lived not within a perimeter but along lines. He calls them "lines of flight," or sometimes "lines of becoming." Such lines prise an opening, even as they bind the animal with its world. Every species, indeed every individual has its own particular line, or rather bundle of lines (Deleuze and Guattari 2004, pp. 224–225). Critically, however, *these lines do not connect*:

> "A line of becoming is not defined by the points it connects, or by the points that compose it; on the contrary, it passes between points, it comes up through the middle, it runs . . . transversally to the localizable relation to distant or contiguous points. A point is always a point of origin. But a line of becoming has neither beginning nor end. . . [It] has only a middle. . . A becoming is always in the middle: one can only get it by the middle. A becoming is neither one nor two, nor the relation of the two; it is the in-between, the . . . line of flight . . . running perpendicular to both" (Deleuze and Guattari 2004, p. 323).

Thus in life as in music or painting, in the movement of becoming – the growth of the organism, the unfolding of the melody, the motion of the brush and its trace – points are not joined so much as swept aside and rendered indiscernible by the current as it flows through. So it is that the line does not link the spider and the fly, or the wasp and the orchid, but "passes between them, carrying them away in a shared proximity in which the discernibility of points disappears" (Deleuze and Guattari 2004, p. 324). Life is open-ended: its impulse is not to reach a terminus but to keep on going. The spider spinning its web or the musician launching into the melody "hazards an improvisation." But to improvise, Deleuze continues, is "to join with the World, or meld with it. One ventures from home on the thread of a tune" (Deleuze and Guattari 2004, pp. 343–344).

If the individual organism is to be understood as a bundle of lines, or what Deleuze calls a *haecceity* (Deleuze and Guattari 2004, p. 290), then what becomes of our original concept of "the environment?" Literally, an environment is that which *surrounds* the organism, yet you cannot surround a bundle without wrapping it up, converting the very paths along which life is lived into boundaries within which it is contained (Ingold 2006, p. 13). Instead, let us imagine ourselves, as did Charles Darwin in *The Origin of Species*, standing before "the plants and bushes clothing an entangled bank" (Darwin 1950, p. 64). Observe how the fibrous bundles comprising every plant and bush are entwined with one another so as to form a dense mat of vegetation. What we have been used to calling the environment reappears on the bank as an immense tangle of lines. Precisely such a view was advanced by the great Swedish geographer, Torsten Hägerstrand, who imagined every constituent of the environment – including "humans, plants, animals and things all at once" – as having a continuous trajectory of becoming. As they move through time and encounter one another, the trajectories of diverse constituents are bundled together in diverse combinations. "Seen from within," wrote Hägerstrand, "one could think of the tips of trajectories as sometimes being pushed forward by forces behind and besides and sometimes having eyes looking around and arms reaching out, at every moment asking "what shall I do next"?" The entwining of these ever-extending trajectories, in Hägerstrand's terms, comprises the texture of the world – the "big tapestry of Nature which history is weaving" (Hägerstrand 1976, p. 332). In this tapestry there are no insides or outsides, no enclosures or disclosures, only openings and ways through. Like Darwin's entangled bank, Hägerstrand's tapestry is a field not of interconnected points but of interwoven lines, not a network but what I shall call a *meshwork* (Ingold 2007, pp. 80–82).

6 Bruno Latour and the Actor Network

I have borrowed the term meshwork from the philosophy of Henri Lefebvre (1991, pp. 117–118). There is something in common, Lefebvre observes, between the way in which words are inscribed upon a page of writing and the way in which the movements and rhythms of human and non-human activity are registered in lived space, but only if we think of writing not as a verbal composition but as a tissue of lines – not as *text* but as *texture*. "Practical activity writes on nature," he remarks, "in a scrawling hand." Think of the reticular trails left by people and animals as they go about their business around the house, village and town. Caught up in these multiple entanglements, every monument or building, viewed in its context and surroundings, is more "archi-textural" than architectural (Lefebvre 1991, p. 118). It, too, despite its apparent solidity and permanence, is a haecceity, experienced processionally in the vistas, occlusions and transitions that unfold along the myriad pathways inhabitants take, from room to room and in and out of doors, as they go about their daily tasks. Like the environment of which it forms a part, the building neither encloses the inhabitant nor is it disclosed from within. "The significant division," as I have argued elsewhere, "is not so much between inside and outside, as between the movement "from the inside going out", and "from the outside going in" " (Ingold 2004, p. 239). As the life of inhabitants overflows into gardens and streets, fields and forests, so the world pours into the building, giving rise to characteristic echoes of reverberation and patterns of light and shade. It is in these flows and counter-flows, winding through or amidst without beginning or end, and not as connected entities bounded either from within or without, that living beings are instantiated in the world.

The distinction between the lines of flow of the meshwork and the lines of connection of the network is critical. Yet it has been persistently obscured, above all in the recent elaboration of what has come to be known, rather unfortunately, as "actor-network theory." The theory has its roots not in thinking about the environment but in the sociological study of science and technology. In this latter field, much of its appeal comes from its promise to describe interactions among people (such as scientists and engineers) and the objects with which they deal (such as in the laboratory) in a way that does not concentrate mind or agency in human hands but rather takes it to be distributed around all the elements that are connected or mutually implicated in a field of action. The term actor-network, however, first entered the Anglophone literature as a translation from the French *acteur réseau*. And as one of its leading proponents – Bruno Latour – has observed in hindsight, the translation gave it a significance that was never intended. In popular usage, inflected by innovations in information and communications technology, the defining attribute of the network is connectivity: "transport *without* deformation, an instantaneous, unmediated access to every piece of information" (Latour 1999, p. 15). But *réseau* can refer just as well to netting as to network – to woven fabric, the tracery of lace, the plexus of the nervous system or the web of the spider. The lines of the spider's web, for example, quite unlike those of the communications network, do not connect points or join things up. They are rather spun from materials exuded from the spider's body and are laid down as it moves about. In that sense they are extensions of the spider's very being as it trails into the environment. They are the lines along which it lives and conduct its perception and action in the world.

The *acteur réseau* was intended by its originators (if not by those who have been beguiled by its translation as network) to be comprised of just such lines of becoming. Their inspiration came, in large measure, from the philosophy of Deleuze. As we have already seen, with acknowledgement to Deleuze, the line of the web does not link the spider to the fly, neither does the latter's line of flight link it to the spider. Ensconced at the centre of its web, the spider knows that a fly has landed somewhere on the outer margins, as it sends vibrations down the threads that are picked up by the spider's super-sensitive, spindly legs. And it can then run along the lines of the web to retrieve its prey. Thus the thread-lines of the web lay down the *conditions of possibility* for the spider to interact with the fly, but they are not themselves lines of interaction. If these lines are relations, then they are relations not *between* but *along*. Of course, as with the spider, the lives of organisms generally extend along not one but multiple lines, knotted together at the centre but trailing innumerable loose ends at the periphery. Thus each should be pictured, as Latour has latterly suggested, in the shape of a star "with a center surrounded by many radiating lines, with all sorts of tiny conduits leading to and fro" (Latour 2005, p. 177). No longer a self-contained object like a ball that can propel itself from place to place, the organism now appears as an ever ramifying web of lines of growth. This is the Deleuzeian *haecceity*, famously compared to a rhizome (Deleuze and Guattari 2004, p. 290). I personally prefer the image of the fungal mycelium. Indeed as the mycologist Alan Rayner (1997) has suggested, the whole of biology would be different had it taken the mycelium as the prototypical exemplar of the living organism. For it could not, then, have been built upon the presumption that life is contained within the absolute bounds of fixed forms. We would rather have a biology that starts from the fluid character of the life process, wherein boundaries are sustained only thanks to the continual flow of materials across them (see also Pearson 1999, pp. 166–168).

7 Ending with Fluid Space

In the science of mind, the absoluteness of the boundary between organism and environment has not gone unquestioned. Thus in a lecture delivered in 1970, the anthropologist Gregory Bateson declared that "the mental world – the mind – the world of information processing – is not limited by the skin" (Bateson 1973, p. 429). His point was that the processing loops involved in perception and action are not interior to the creature whose mind we are talking about, whether human or non-human, nor can that creature's activity be understood as the merely mechanical output of one or more cognitive devices located in the head. Rather, such activity has to be understood as one aspect of the unfolding of a total system of relations comprised by the creature's embodied presence in a specific environment. Much more recently, in his book *Being There*, Andy Clark has made the same point. The mind, Clark tells us, is a "leaky organ" that refuses to be confined within the skull but mingles shamelessly with the body and the world in the conduct of its operations (Clark 1997, p. 53). More strictly, he should have said that the skull is leaky, whereas the mind is what leaks! From Bateson to Clark, however, there remains a presumption that whereas the mind leaks, the organism does not. Whatever we might say about the mind, and about its propensity to mingle with the world along the multiple pathways of sensory engagement with its surroundings, the organism at least remains confined within the envelope of the body. This presumption, along with the division between mental and organic activity on which it rests, seems to me to be unsustainable. For how can there be any sensory engagement that does not also involve a flow of materials within a wider field of forces? For this reason I would like to return to Bateson's declaration and take it one step further. I want to suggest that, as a nexus of life and growth within a meshwork of relations, *the organism is not limited by the skin*. It, too, leaks.

Another way to express this is to say that organisms inhabit what Annemarie Mol and John Law (1994) have called "fluid space." In fluid space there are no well-defined objects or entities. There are rather substances which flow, mix and mutate, sometimes congealing into more or less ephemeral forms that can nevertheless dissolve or re-form without breach of continuity (Mol and Law 1994, pp. 659–664). Every line – every relation – in fluid space is a path of flow, like the riverbed or the veins and capillaries of the body. As the sanguinary image suggests, the living organism is not just one but a whole bundle of such lines. In a quite material sense, lines are what organisms are made of. Indeed anatomists have always known this as they have spoken of bodily "tissues" (Ingold 2007, p. 61). For the tissue is a texture formed of myriad fine threads tightly interlaced, presenting all the appearance, to a casual observer, of a coherent, continuous surface. To the anatomical gaze, however, the organic tissue becomes – as J. Arthur Thomson wrote in 1911 – "in a quite remarkable way translucent," resolving into its constituent threads of nerve, muscle, blood-vessels and so on (Thomson 1911, p. 27). What is the nervous system, asked the philosopher Henri Bergson, if not "an enormous number of threads which stretch from the periphery to the centre, and from the centre to the periphery" (Bergson 1991, p. 45)? Indeed the skin is not an impermeable boundary but a permeable zone of intermingling and admixture, where traces can reappear as threads and vice versa (Ingold 2007, pp. 59–61). It is not, then, that organisms are entangled in relations. Rather, every living thing is itself an entanglement, a tissue of knots whose constituent strands, as they become tied up with other strands, in other bundles, comprise the meshwork.

To appreciate the distance we have travelled, let me return in conclusion to Gibson. Recall his assumption that to render the open world habitable, it must first be furnished with *objects*, and thus partially enclosed. Now it is by their outward surfaces, according to Gibson, that objects are revealed to perception. It is as though they had turned their backs on inhabitants, exposing their congealed shapes and layouts, rather than allowing inhabitants to join with them in the material flows and movements contributing to their ongoing formation. Every surface, as Gibson explains, is an interface between the more or less solid substance of an object and the volatile medium that surrounds it. If the substance is dissolved or evaporates into the medium, then the surface disappears and with it the object it once enveloped (Gibson 1986, pp. 223). Thus the very objectness of things lies in the separation and immiscibility of substance and medium. Remove every object, however, and a surface still remains – for Gibson the most fundamental surface of all – namely the ground, marking the interface between the substance of the earth below and the gaseous medium of the sky above (Gibson 1986, pp. 10, 33). Has the earth, then, turned its back on the sky? If it had, then as Gibson correctly surmised, no life would be possible. The open could not be inhabited. Our conclusion, to the contrary, is that the open *can* be inhabited precisely because, wherever life is going on, the interfacial separation of earth and sky gives way to mutual permeability and binding. For what we vaguely call the ground is not, in truth, a coherent surface at all but – just like the skin – a zone in which the air and moisture of the sky combine with substances whose source lies in the earth in the ongoing formation and growth of living organisms (Ingold 2008, pp. 30–1).

Thus, far from inhabiting a sealed ground furnished with objects, the animal lives and breathes in a world of earth and sky – or becoming earth and becoming sky – where to perceive is to align one's movements in counterpoint to the modulations of day and night, sunlight and shade, wind and weather. It is to feel the currents of air as it infuses the body, and the textures of the earth beneath one's feet. In the open world, to leave the last word to Deleuze, "there is no line separating earth and sky; there is no intermediate distance, no perspective or contour, visibility is limited; and yet there is an extraordinarily fine topology that relies not on points or objects but rather on haecceities, on sets of relations (winds, undulations of snow or sand, the song of the sand or the creaking of the ice, the tactile qualities of both)" (Deleuze and Guattari 2004, p. 421). These haecceities are not *what* we perceive, since in the world of fluid space there are no objects of perception. They are rather what we perceive *in*. In short, to perceive the environment is not to take stock of its contents but to *follow what is going on*, tracing the paths of the world's becoming, wherever they may lead us.

References

Agamben G (2004) The open: man and animal. Translated by K Attell. Stanford, CA: Stanford University Press.

Bateson G (1973) Steps to an ecology of mind. London: Granada.

Bergson H (1991) Matter and memory. Translated by NM Paul and WS Palmer. New York: Zone Books.
Clark A (1997) Being there: putting brain, body and the world together again. Cambridge, MA: MIT Press.
Darwin C (1950) On the origin of species by means of natural selection, or, the preservation of favoured races in the struggle for life. London: Watts [reprint of First Edition of 1859].
Deleuze G, Guattari F (2004) A thousand plateaus: capitalism and schizophrenia. Translated by B Massumi. London: Continuum [originally published in 1980].
Dreyfus HL (1991) Being-in-the-world: a commentary on Heidegger's 'Being and Time, Division I'. Cambridge, MA: MIT Press.
Elden S (2006) Heidegger's animals. Cont Phil Rev 39: 273–91.
Geertz C (1973) The interpretation of cultures. New York: Basic Books.
Gibson JJ (1986) The ecological approach to visual perception. Hillsdale, NJ: Lawrence Erlbaum [originally published 1979].
Griffin DR (1984) Animal thinking. Cambridge, MA: Harvard University Press.
Hägerstrand T (1976) Geography and the study of the interaction between nature and society. Geoforum 7: 329–34.
Harrison P (2007) The space between us: opening remarks on the concept of dwelling. Environ Planning D: Soc Space 25: 625–647.
Heidegger M (1995) The fundamental concepts of metaphysics: world, finitude, solitude. Translated by W McNeil, N Walker. Bloomington, IN: Indiana University Press [based on a course presented in 1929–30, originally published in 1983].
Ingold T (1989) The social and environmental relations of human beings and other animals. In: Standen V, Foley RA (eds) Comparative socioecology. Oxford: Blackwell Scientific, pp. 495–512.
Ingold T (1992) Culture and the perception of the environment. In: Croll E, Parkin D (eds) Bush base: forest farm. Culture, environment and development. London: Routledge, pp. 39–56.
Ingold T (2000) The perception of the environment: essays on livelihood dwelling and skill. London: Routledge.
Ingold T (2004) Buildings. In: Harrison S, Pile S, Thrift N (eds) Patterned ground: entanglements of nature and culure. London: Reaktion, pp. 238–40.
Ingold T (2006) Rethinking the animate, re-animating thought. Ethnos 71(1): 1–12.
Ingold T (2007) Lines: a brief history. London: Routledge.
Ingold T (2008) Earth, sky, wind and weather. In: Hsu E, Low C (eds) Wind, life, health: anthropological and historical perspectives. Oxford: Blackwell, pp. 17–35.
Latour B (1999) On recalling ANT. In: Law J, Hassard J (eds) Actor network theory and after. Oxford: Blackwell, pp. 15–25.
Latour B (2005) Reassembling the social: an introduction to actor-network theory. Oxford: Oxford University Press.
Lefebvre H (1991) The production of space. Translated by D Nicholson-Smith. Oxford: Blackwell [originally published in1974].
Lewontin RC (1982) Organism and environment. In: Plotkin HC (ed) Learning, development and culture. Chichester: Wiley.
Mol A, Law J (1994) Regions, networks and fluids: anaemia and social topology. Social Studies Sci 24: 641–71.
Pearson KA (1999) Germinal life: the difference and repetition of Deleuze. London: Routledge.
Rayner ADM (1997) Degrees of freedom: living in dynamic boundaries. London: Imperial College Press.
Sahlins M (1976) Culture and practical reason. Chicago: University of Chicago Press.
Thomson JA (1911) Introduction to science. London: Williams and Norgate.
von Uexküll J (1982) The theory of meaning. Translated by B Stone and H Weiner from Bedeutungslehre (ed. T. von Uexküll). Semiotica 42(1): 25–82 [originally published in 1940].
von Uexküll J (1992) A stroll through the worlds of animals and men: a picture book of invisible worlds. Semiotica 89(4): 319–391 [originally published in 1934].
Whitehead AN (1938) Science and the modern world. Harmondsworth: Penguin.

Index

Printing: Krips bv, Meppel, The Netherlands
Binding: Stürtz, Würzburg, Germany